Studies in Systems, Decision and Control

Volume 168

Series editor

Janusz Kacprzyk, Polish Academy of Sciences, Warsaw, Poland
e-mail: kacprzyk@ibspan.waw.pl

The series "Studies in Systems, Decision and Control" (SSDC) covers both new developments and advances, as well as the state of the art, in the various areas of broadly perceived systems, decision making and control–quickly, up to date and with a high quality. The intent is to cover the theory, applications, and perspectives on the state of the art and future developments relevant to systems, decision making, control, complex processes and related areas, as embedded in the fields of engineering, computer science, physics, economics, social and life sciences, as well as the paradigms and methodologies behind them. The series contains monographs, textbooks, lecture notes and edited volumes in systems, decision making and control spanning the areas of Cyber-Physical Systems, Autonomous Systems, Sensor Networks, Control Systems, Energy Systems, Automotive Systems, Biological Systems, Vehicular Networking and Connected Vehicles, Aerospace Systems, Automation, Manufacturing, Smart Grids, Nonlinear Systems, Power Systems, Robotics, Social Systems, Economic Systems and other. Of particular value to both the contributors and the readership are the short publication timeframe and the world-wide distribution and exposure which enable both a wide and rapid dissemination of research output.

More information about this series at http://www.springer.com/series/13304

Václav Talhofer · Šárka Hošková-Mayerová
Alois Hofmann

Quality of Spatial Data in Command and Control System

 Springer

Václav Talhofer
Faculty of Military Technology, Department
  of Military Geography and Meteorology
University of Defence
Brno, Czech Republic

Alois Hofmann
Faculty of Military Technology, Department
  of Military Geography and Meteorology
University of Defence
Brno, Czech Republic

Šárka Hošková-Mayerová
Faculty of Military Technology,
  Department of Mathematics and Physics
University of Defence
Brno, Czech Republic

Reviewed by: Prof. Wolfgang Reinhardt, University of Bundeswehr, Munich, Germany,

and

Dr. Marie Vojtíšková, Military Geographic and Hydrometeorological Institute, Dobruška, Czech Republic.

ISSN 2198-4182   ISSN 2198-4190   (electronic)
Studies in Systems, Decision and Control
ISBN 978-3-030-06876-9   ISBN 978-3-319-94562-0   (eBook)
https://doi.org/10.1007/978-3-319-94562-0

Printed on acid-free paper

This Springer imprint is published by the registered company Springer Nature Switzerland AG
The registered company address is: Gewerbestrasse 11, 6330 Cham, Switzerland

Contents

About the Authors

Václav Talhofer is Full Professor of Cartography and Geoinformatics at the University of Defense in Brno, Czech Republic.

Šárka Hošková-Mayerová is Associate Professor of Mathematics at the University of Defense in Brno, Czech Republic.

Alois Hofmann is a teacher and scientist of Cartography and Geoinformatics at the University of Defense in Brno, Czech Republic.

Abbreviations

2D	Two-Dimensional
3D	Three-Dimensional
ACC	Accuracy Code
ACR	Army of the Czech Republic
AOI	Area of Interest
AOR	Area of Responsibility
C2S	Command and Control System
CIT	Communication and Information Technology
CTO 25	Catalogue of Topographic Objects
CZUK	State Administration of Land Surveying and Cadastre
DGI	Digital Geographic Information
DGIWG	Defense Geographic Information Working Group
DMR3	Digital Model of Relief of the Czech Republic of the 3th Generation
DMR4	Digital Model of Relief of the Czech Republic of the 4th Generation
DMR5	Digital Model of Relief of the Czech Republic of the 5th Generation
DMU100	Digital Model of Territory 100
DMU25	Digital Model of Territory 25
DoD	Department of Defense (USA)
EGM96	Earth Gravitation Model 1996
ERDC	Engineer Research and Development Center
GCS	Geodetic Coordinate System
GD	Geographic Data
GEOINT	Geospatial Intelligence
GI	Geographic Information
GIS	Geographic Information System
GNSS	Global Navigation Satellite System
ICA	International Cartographic Association
IDW	Inverse Weight Distance tool
IMIT	Imagery Intelligence
INSPIRE	Infrastructure for Spatial Information in Europe

ISO	International Organization for Standardization
ISO/TC	ISO/Technical Committee
IUV	Individual Utility Value
JRC	Join Research Center
KML	Keyhole Markup Language
LCC	Lambert Conformal Conic
LOD	Level of Detail
LUC	Land Use Code
MCA	Multi-criteria Decisional Analysis
MGCP	Multinational Geospatial Co-Production Program
MTA	Military Training Area
NATO	North Atlantic Treaty Organization
NRMM	NATO Reference Mobility Model
OGC	Open GIS Consortium
PDF	Probability Distribution Function
RCE	Relative Cost Effectiveness
RH	Real Hours
SD	Soil Database
SRTM	Shuttle Radar Topographic Mission
TIN	Triangular Irregular Network
TTD	Tactical–Technical Data
UM	Units of Measure
USA	United States of America
UTM	Universal Transverse Mercator
VCI	Vehicle Cone Index
WGS84	World Geodetic System 1984

Introduction

Nowadays, human society is rather dependent on information technology. Basically, anywhere in the world, information is available from any place on Earth, including geographic information. We can find very detailed data about any area in the world at geoinformation portals such as Google Maps or Bing Maps. The amount and availability of digital geographic information is no longer a major issue for most of the professions that work with it. Most spatial information users welcome this fact, and spatial information is becoming an integral part of their everyday life, both professional and private one.

However, what is becoming the key issue is the assurance of quality of this information. A typical user does not directly notice or deal with the issues of quality of spatial data, although poor-quality and unreliable information can cause problems, for example, when using a database in a satellite navigation that has not been updated, the navigation can direct them to a wrong location. If such a situation happens on a holiday journey, it is often only a mere inconvenience. However, if it happens during an ambulance journey, the life of a patient may be endangered.

This monograph aims to familiarize readers with the issue of evaluating the quality of digital geographic information in terms of their use. The authors draw on their long-term experience of working with digital geographic information in the army environment as well as in the services of the integrated rescue system in the Czech Republic. It tries to identify the key requirements for the functionality of this information and describes the system of evaluation of its quality. The entire text is supplemented with examples, which document the effect of varying quality of information on the whole decision-making process in command and control systems at the level of the army and rescue services. The monograph presents the results of the research supported by the Ministry of Defence of the Czech Republic-project NATURENVIR.

This monograph is primarily intended for professionals responsible for the implementation of digital geographic information into command and control systems, or alternatively for professionals who use these systems in their work. For this reason, particular attention is paid to user aspects.

The authors thank reviewers—Prof. Wolfgang Reinhardt from University of Bundeswehr, Munich, Germany, and Dr. Marie Vojtíšková from Military Geographic and Hydrometeorological Institute, Dobruška, Czech Republic—for the valuable remarks and kind comments.

Václav Talhofer
Šárka Hošková-Mayerová
Alois Hofmann

Chapter 1
Geographic Support of Decision-Making Processes

The current way of controlling operational units moving in a given space cannot work without a well-thought way of command and control. *Command and control* may be generally defined as a system consisting of personal, technical, and legislative components created to provide coordination and commanding of operational units with the aim to complete required goals of an operation. A significant part of command and control is played by *the decision-making process*, during which an individual or a team chooses the best solution of the all possible ones. Depending on the task, facts about the given task are found, as well as connections with solution of other tasks, factors that influence the given task, and preferences that need to be considered before making the decision.

1.1 Used Terminology

Decision-making processes may be intuitive or structured. Intuitive decision-making processes are rather the domain of the personality approach, in which it is basically not necessary to stick to the established procedures and rules. Within the system of command and control, however, decision-making processes are mostly based on multi-criteria analyses (Maliene, 2011), i.e. as the structured.

If the decision-making processes concern solving tasks in a real geographic space, they cannot work without relevant information about this space. The most common form of such information is *geographic information* (GI) obtained mostly from *geographic data* (GD), that enables subjects of decision-making processes (commanders and staff) to recognize the environment in which the given tasks will be solved, i.e. to discover facts about this environment. Generally, they also enable to assess *geographic factors*, i.e. to judge the influence of geographic environment on the solution of the given task.

Geographic information may be in paper or digital forms, i.e. in the form of maps and plans, or in the form of *digital geographic information* (DGI). All forms of

© Springer International Publishing AG, part of Springer Nature 2019   1
V. Talhofer et al., *Quality of Spatial Data in Command and Control System*, Studies
in Systems, Decision and Control 168, https://doi.org/10.1007/978-3-319-94562-0_1

geographic information for command and control are usually provided by organizations (state as well as commercial) working apart from command and control itself. However, a stable and resistant system of command and control needs a permanent access to geographic information and possibly also to technical help while worked with. Such access to geographic information in the process of command and control is called *geographic support*.

The resulting quality of the decision depends also on the quality and reliability of geographic data and information. In general, these data and information can be considered to be a product whose production involves many workers of different professions and various technological procedures. It is therefore possible to evaluate these data and information based on its quality as other engineering products. Evaluation of the quality and reliability is dealt with by both national and international organizations, which define their own concepts of quality, as well as their evaluation methods (ESRA, 2017), (Vintr & Vintr, 2013). Using the definition of quality from the above sources, it is possible to generally define the quality of geographic information as follows: "*Geographic information entering the system of geographic support of decision-making processes is considered to be of quality if it provides information about geographic environment from the place where the task is fulfilled and in such a level of detail and currency that corresponds to the way of solving the task. Information is provided to the place where the decision-making processes take place in time, in the required form and with the required functionality.*"

1.2   Ways of Command and Control, Modelling of Actions, Process Modelling

According to Miklošík (1997), in armies of the 2nd half of 18th century firearms of a longer fire range started to be used massively which led to changes in combat tactics. Irregular combat sets started to be used and combat spread also into an unclear and indented terrain where armies could cover against the effects of firearms. From the point of view of organization of units on the battlefield and the command of armies, it was necessary to use more of the characteristics of the terrain, topographic map of which proved to be very suitable. Topographic maps gradually evolved in a way that enabled them to meet the ever-growing requirements for the content quality, which was meant to ensure their usability both during planning a battle and commanding units on a field.

This historical experience manifests also in present and not only in the field of military, but also in the civilian environment where it is necessary to plan and coordinate actions of a large number of active elements (units, groups) on a large area. Classic paper topographic maps were at the end of the 20th century gradually supplemented and partially substituted with digital information about the environment, usually in the form of digital geographic data. From the point of view of the complex understanding of the complete support system of decision-making based on spatial data (so-called *geographic support*, here also meant as *Geospatial intelligence support, GEOINT*), for non-military as well as military actions it is possible to follow the material by the Department of Defence of the USA (DoD) Geospatial Intelligence in Joint Operations (Staff, 2012). According to the mentioned publication "Geospatial intelligence supports user requirement by providing geo-referenced visual and data products (e.g., maps, charts, digital file, imagery, and digital raster or vector information) that serve as a foundation and common frame of reference for any joint operation. GEOINT operations are the tasks, activities, and events to collect, manage, analyze, generate, visualize, and provide imagery, IMINT, and geospatial information necessary to support national and defense missions and international arrangements. GEOINT operations are comprised of a set of interrelated and specific activities and procedures to conduct GEOINT and cross-functional operational awareness of the environment. These activities continuously support information fusion, visualization, analysis, and sharing." Similarly, we can define aims and goals of geographic support basically of all modern armies as well as units of integrated rescue systems.

In *command and control systems* (C2S) *in rescue systems, armed forces and the police* mainly data produced by state mapping agencies are used. The reason for that is the necessity to secure permanent quality of the data and the certainty of their permanent update. Positional and elevation data are often also the background for adding thematic data from the individual units. For instance, in armed forces they are kinds of armies for which thematic add-ons such as operational-tactical add-on, data of artillery, engineers, air defense, etc. are being developed. All data that is used by most of the armed forces of NATO states are localized in position in the *geodetic coordinate system* (GCS) WGS84 (Word Geodetic System 1984) and usually in UTM projection (Universal Transverse Mercator), with the exception of data for air force that is in LCC projection (Lambert Conformal Conic) which corresponds to the approved standards within NATO. Basic source data usually have pre-declared accuracy given by the class of accuracy which agencies guarantee by the system of output inspection. Thematic data of units that are in the command and control systems integrated with source data rarely have the declared characteristics of accuracy, and their positional and thematic accuracy can only be estimated based on procedures that are used for their collection and processing. As the source of data here, either technical sources (observation sensors such as radars or passive lasers, optical systems) are used here or the data are measured in a terrain using similar procedures to mapping technology, nowadays mainly system in the *global navigation satellite systems* (GNSS). Many thematic data that C2S work with are localized without the contact with terrain, only based on the source situation. This is especially localization of an operational situation, a description of places of events (traffic accidents, criminal action, natural disasters, etc.). C2S often also use sources of field thematic data and information outside the area of geographic and topographic geoinformatics (Konecny & Reinhardt, 2010). They are for instance data and information about soils, vegetation, hydrology, etc. The mentioned sources are, however, not completely described, their qualitative characteristics usual with source topographic databases is not known. Nevertheless, they are often worked with the same way as with source data with known characteristics.

1.3   Spatial Analyses of Data as Source of Information for C2S

Digital geographic data after their visualization may be used basically in the same way as classic paper maps, getting all necessary information only by studying the visualized image and making the desired analyses of content exclusively in a user's head. Such use of digital data would be absolutely inappropriate and uneconomical. The strength of digital geographic data together with tools for working with them lies in the possibility to make spatial geographic analyses in which influences of geographic environment on the intended or realized activity are evaluated. Spatial geographic analyses are mostly created based on a general requirement arising from the occurred situation, or from the superior, commander, or their staff. The aim of such analysis is to recognize concrete current influences of geographic environment and current weather conditions on the ongoing or planned activities (Kovarik & Rybansky, 2014).

The resulting value of the analysis depends on many factors. They are especially the quality of the source data, quality and trustworthiness of the used mathematical and information model, and eventually also the level of knowledge and competences of the analyst who makes the particular analysis. The overall parameters of the quality of data, models, and analysts may significantly influence the reached results. Yet, most of created analyses are provided to the ordering party without giving them any information about the quality of the source data and its influence on the reliability of the resulting analysis. This situation is documented in the picture (Fig. 1.1).

From the point of view of using the results of an analysis, this usual procedure offers several advantages, but also disadvantages.

The advantages are:

– Relatively simple solution of a task,
– Usually clear result,
– The ordering party does not have to consider the reliability characteristics of the reached result.

The disadvantages of the stated solution are especially:

– Total dependence on spatial data without considering their quality,
– Not knowing the quality of the results of the analysis,
– As the ordering party does not have additional information about the quality of the required result, they are left to use only one solution and do not have the possibility of choice between more options.

To minimize the above-mentioned disadvantages, it is possible to involve also an evaluation of quality of the source data in results of spatial analyses.

If also the quality of spatial data is taken into consideration in the process of spatial analysis, the geographer-analyst may create more options of the solution depending on how they handle its characteristics. It is possible to make options that are reliable, less reliable, or considerably risky. The ordering party then gets not only the result

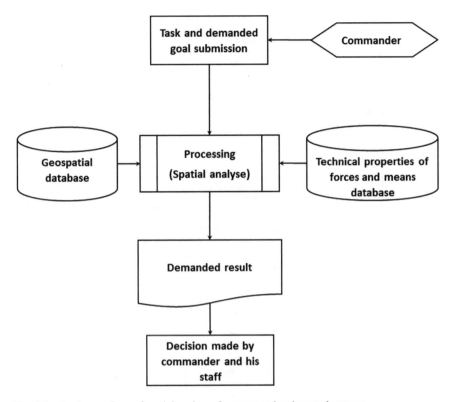

Fig. 1.1 Usual way of use of spatial analyses for command and control systems

itself, but also other information about the level of risk of the solution regarding the quality of the used data. They may get the information about the quality of the result as an answer to their direct question, or, more often in distributed systems, as visualized information with the help of the above-mentioned methods. It is then their responsibility what other steps to make. Either they are satisfied with all received information or instruct the geographer to improve the quality of the source data and make a new version of the given spatial analysis. For doing that, the geographer must get corresponding conditions for improvement of this quality—time, staff and material. The model (Fig. 1.2) illustrates this procedure.

1.4   User's Functions of Digital Geographic Data and Information

DGI may be used similarly to classic analogue background—maps that are displayed on screens of the stated systems. Their main domain, however, is their use directly in the digital form, e.g. as source data for making various types of analyses, for con-

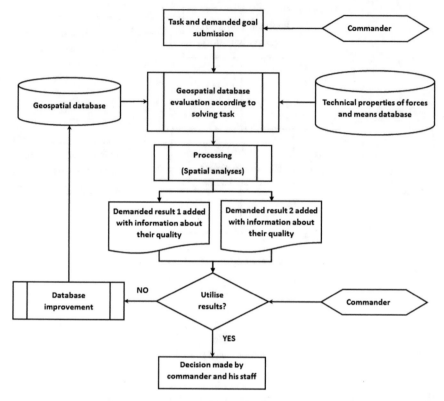

Fig. 1.2 Process of creation of spatial analyses with consideration of the quality of the source data

trolling actions, making prognosis, etc. The range of use of DGI is significant and keeps getting wider. The ways of usage of DGI projected into user's requirements on the content and data quality is the decisive moment for organizations and institutions that provide the data. From a huge and fragmented set of users' tasks it is difficult to define general users' requirements on DGI. That is why it is necessary with certain generalization to find common points of all activities and summarize them into general types of tasks and ways of usage of DGI in C2S. Based on this, it is possible to define six basic functions that cover their main features (Talhofer, 2004):

1. *Information function* that expresses the ability of DGI to quickly and reliably provide information about position and basic characteristics saved in topographic and geographic objects and phenomena in the area of interest. However, it does not have to be only providing information about the individual objects and phenomena, but also information about the complete characteristics of the area of interest, such as for instance the structure of surface, character of soil and vegetation cover, demographic characteristics, etc.
2. *Model function* expressing the ability of DGI to serve as a model for deriving geometric and other relationships among topographic and other objects and phe-

nomena and their characteristics. With the help of this function it is possible to find geometric characteristics of individual objects and phenomena (length, width, circumference, area, volume, …), their topologic relationships (neighborhood, connection, crossing, overlays, …), and relationships between their attributes. The use of this function may bring completely new information and findings about the area of interest that manifest only in a suitable combination of analysis of the individual saved objects and phenomena (e.g. analysis of visibility from a given place of observation when considering the relief of terrain as well as its cover).

3. *Function of the background for mathematical modelling, projecting and planning* that applies in cases when with the help of DGI we create an intention of a future activity or work that is being projected and that only will be realized in the future. This function also applies in cases when possible consequence of realization of a future state are studied, both planned (e.g. construction of roads and their influence on the surrounding environment) or unplanned (e.g. possible consequences of effects of a tidal wave in case of a dam bursting, level of danger of an area with a potential avalanche, etc.). This function enables to optimize projected and planned intentions beforehand in a way so that their effects are according to their purpose either maximized, or on the contrary minimized. For realization of the given function we use methods of mathematical modelling.

4. *Function of automation* for controlling of realization process of projected and planned intentions. This function applies for instance during orientation on the move (on land as well as in air), during coordination or observation of movement of a higher number of objects (air traffic control, observation of situation in air in a defined region, observation of movement of means of transport, etc.) or when controlling and observing the course of construction of roads and other constructions.

5. *Illustrational function*, which expresses the ability of DGI to serve for illustration of a situation, passing messages with means of communication systems, etc. Condition for its usage is a suitable visualization of the saved data, which basically must correspond to the principles of cartographic semiotics. Data may be visualized on screens of computers, tablets, mobile phones, etc., displayed by projectors, or printed on a suitable medium (paper, foil, etc.). For visualization characteristics of the graphic form of information is used. They are clear, illustrative and unambiguous.

6. *Function of source* for deriving other types of *geographic information systems* (GIS) and maps and for cartographic purposes. Databases DGI may hide almost "Unlimited" number of data. It is limited basically only by the size of an allocated memory space for their storing and possibilities of the software to work with the given volume of data and solve tasks in the required time. From this data it is possible to derive other databases by simple or conditional selections; or use them for cartographic creation supported by information technology, i.e. for the creation of topographic as well as a wide range of special (thematic) maps.

Such defined functions form the basis for assessment of utility characteristics of the given database DGI from the point of view of its usage for a specific task or a set of tasks. The assessment of user characteristics is discussed in the Chap. 2.

References

ESRA. (2017, 12). *European safety and reliability association*. Retrieved from ESRA—European Safety and Reliability Association: http://www.esrahomepage.eu/home.aspx?lan=230&tab=276 4&itm=2764&pag=1625.

Konecny, M., & Reinhardt, W. (2010). Early warning and disaster management: The importance of geographic information (Part A). *International Journal of Digital Earth, 3*(3), 217–220. https://doi.org/10.1080/17538947.2010.508884.

Kovarik, V., & Rybansky, M. (2014). Selecting locations for landing of various formations of helicopters using spatial modelling. In *8th International Symposium of the Digital Earth (ISDE8) 2013*, (Vol. 18, p. 6). Kuching, Malajsie: Institute of Physics Publishing (IOP).

Maliene, V. (2011, May). Specialised property valuation: Multiple criteria decision analysis. *Journal of Retail & Leisure Property, 9*(5), 443–450.

Miklošík, F. (1997). *State maps in the Czech Republic (Státní mapová díla v České republice, In Czech)*. Brno: Vojenská akademie v Brně.

Staff, U. A. (2012). *Geospatial inteligence in joint operations* (Joint Publication 2–03 ed.). Whasington, D.C., USA: Joint Chiefs of Staff.

Talhofer, V. (2004). Digital geographic data: Potential evaluation. *AGILE 2004, 7th Conference on Geographic Information Science, Conference proceedings* (pp. 675–686). Heraclion, Crete, Greece: AGILE.

Vintr, Z., & Vintr, M. (2013). Evaluation of product reliability during warranty. *Applied Mechanics and Materials*, 436, 557–562. Zurich: Trans Tech Publications.

Chapter 2
Quality of Digital Geographic Data and Information

Until DGI became widely used, commanders and staff got information about the area of activity first of all from paper models of landscape, i.e. from maps. International Cartographic Association (ICA) defines a map as ICA (2014): "*A map is a symbolized representation of geographic reality, representing selected features or characteristics, resulting from the creative effort of its author's execution of choices, and is designed for use when spatial relationships are of primary relevance.*" The efficiency of decision-making processes using maps has always been significantly influenced by their quality. Quality prepared and up-to-date maps enable to get detailed information about the area of activity that is necessary for making optimal decisions. For the needs of decision-making processes, there are maps published that correspond to the given needs, e.g. units of integrated rescue systems or armed forces, and they are usually produced by state organizations that guarantee its constant quality, reliability and regular updating.

When using paper maps, the factor of user's ability plays an important role. An experienced user is able to correctly understand also such associations between objects and phenomena on land surface that are not in the contents of maps directly recorded, or are recorded wrongly or not accurately. For example, if in a map a symbol of a small bridge on a road is missing and there is a creek on this road, the user will understand this mistake and he will consider the small bridge on the road.

There is a more complicated situation in case of wide usage of DGI. Relatively trouble-free availability of basic as well as thematic geographic data in the environment of *communication and information technology* (CIT), for instance through web services, may lead to an assumption that also in decision-making processes it is possible to use any available geographic data, especially when such data are freely available. The danger of such approach, however, lies especially in the circumstances:

- Not knowing, or little knowledge of the data origin,
- Not knowing, or little knowledge of the system management and updating of data,
- Not knowing, or little knowledge of the technical data quality,
- Permanent reliability is not secured.

© Springer International Publishing AG, part of Springer Nature 2019
V. Talhofer et al., *Quality of Spatial Data in Command and Control System*, Studies in Systems, Decision and Control 168, https://doi.org/10.1007/978-3-319-94562-0_2

– Permanent accessibility is not secured.
– Data sets may have heterogeneous positional and thematic characteristics.
– Data sets do not have to correspond to the required standards.
– Data sets do not have to be usable in the environment CIT in which the decision-making process itself takes place.

Moreover, a typical user of DGI is primarily a system based on communication and information technology whose activity is controlled with the help of software, e.g. automated systems of command and control or navigation systems. Man, as a user, enters the process of working with data mostly at the beginning and then only at its end when he/she gets results of the work of the program system via graphic language.

The manipulation with data is controlled by ready programs or tools. Programs and tools unlike people are much faster, but their algorithms until now do not allow to interpret missing, incomplete or not up-to-date information based on the complete context of the used data. With the growing complexity of programs and work with large files of data, also the danger that the results processed by the system will be wrong is growing. This concerns especially complex spatial analyses. Moreover, to reveal a mistake is usually rather complicated. A faulty result may manifest only when used in practice which in many cases may be too late.

The basis of a successful work with DGI in decision-making processes (but not only in them) lies in thorough knowledge of parameters of quality of these data and acceptance of these parameters in the whole process of working with them. The following text briefly discusses an explanation of the basis of modelling geographic reality and geographic problems, as well as with the quality of digital geographic data.

2.1 Digital Geographic Data

For the needs of work in environment of communication and information systems and technology, geographic area is modelled with the help of models. A model is—generally speaking—a representation of the given geographic reality. With regards to the inner complexity of the world and number of interactions in them, the models are created as a simplified and controlled view of reality. Models may help to understand, describe and expect how things work in the real world (Tomlin, 1990; Fotheringham, 2000; Lloyd, 2011).

In practice there are two main types of models:

– *Representative* models—represent objects in the country,
– *Geo-process* models—simulate processes in the country.

Representative models describe objects in the country, for instance buildings, water courses or vegetation. Representative models depict the geometry of an object (e.g. the shape of a building, block of buildings, water reservoirs, etc.) and at the same time also spatial relationships between other objects in landscape (for instance location of buildings in a certain locality). These models also describe chosen characteristics of objects (e.g. out of which material a building is built, who owns it, etc.). Representative models are often marked as data and they are considered as descriptive models. This data identifies geographic location and characteristics of natural and anthropogenic objects and phenomena and boundaries between them. The data are called spatial data, geospatial data or geographic data. Geographic information is, in fact, the data that the recipient gives some meaning based on the basis of his knowledge and experience and which reduces the entropy (uncertainty) of the recipient, taking into account his needs and requirements (Sklenak, Berka, Rauch, Strossa, & Svatek, 2001). But in practice, spatial data, geographic data and geographic information are often confused these terms are often interchangeable.

Geo-process models depict interaction of objects that are saved in the representative model. Geo-process models are mostly created based on an analysis of behavior of objects within the given problem that is being solved. If geo-process models are created in the environment of GIS, they are usually compiled from a wide range of tools that are a part of the program system GIS. Geo-process models may be used to describe processes but also for prediction of what is going to happen when a certain action takes place.

Geographic support of decision-making processes is based just on geo-process models, which, however, work above DGI creating representative models. That is the reason why the following text is dedicated to these representative models.

2.1.1   Geospatial Data

The aim of creation of a representative data model is to create a basic data structure of objects and phenomena that will be saved in geodatabases and accurately describe their geometric, topological and thematic characteristics.

Basic data structure is created by geographic objects—*entities*, as basic information units that in the given level of detail are not further divided and have invariable geometric, topological and thematic characteristics. For geographic objects it is possible to define the following characteristics more specifically:

– Objects have their own *shape*, i.e. their own *geometry*.
– Objects have *spatial coordinates* that locate them in a given geodetic reference positional and elevation system and possibly also in a given cartographic projection. The location in the geodetic reference system is given usually in geographic coordinates (longitude and latitude). If the system is supposed to work in planar coordinates, it is necessary to transform the geographic coordinates into a plane using a suitable cartographic projection. The plane is then given in Cartesian coordinates or it is expressed by indices of raster or matrix field in this projection. In case objects have complete spatial coordinates, it is necessary to take also geodetic reference elevation system into account. Objects then express also their own eleva-

tion or they are localized in geocentric coordinates X, Y, Z. For further information on the topic, see e.g. Talhofer (2007) or Snyder (1987).

- Objects have *thematic characteristics (attributes)* that can be expressed in a number format, as a text, as a logical expression, as a web link, etc.
- Objects are included into a *higher hierarchical groups*. All objects of the higher hierarchical group have the same given characteristics and differ from each other with other characteristics. E.g. an object of hydrological network is a part of a water course with invariable characteristics, such as its width, depth and speed of water course. Its basic characteristics is that it drains rainfall water from the given area. Individual objects, however, are logically united in water courses which in is this case a higher hierarchical group. Another classification of this hierarchical group may be a division into navigable and unnavigable water courses, or their division into surpassable or unsurpassable with the given equipment, etc.
- Objects have *spatial relationships* to their neighbors, i.e. *topological characteristics*.
- Objects can have also *non-spatial relationships* to other objects, e.g. relationship between a house and its owner.
- Objects usually have *only chosen characteristics* that may be considered. This fact can be used with a definition of data model with the aim to prevent—during putting information into the database—stating characteristics that the given object may never have, or enabling to put in only such characteristics that the object may have.
- E.g. an object *part of railway* cannot have characteristics *number of inhabitants*, also characteristics *electrification* may have only values *unknown, electrified, not electrified, others* and so on.
- Objects must obey certain *rules*. These rules may be included when defining behavior of an object and when filling in information. For instance, power lines cannot be connected with gas lines or water pipelines.

The stated characteristics concern models of real geographic objects, however, in the same way it is possible to approach also objects modelling natural as well as social phenomena. An example of a natural phenomenon is e.g. phenomena in the atmosphere (model of pressure or thermal field). With social phenomena it may be e.g. the place of a police unit action with its characteristics, area of an increased risk, etc.

From the point of view of geometric characteristics, objects are defined as:

- *Points* represent objects whose dimensions are so small that it is not possible to record them as areas in the given level of detail.
- *Lines* are used for representation of objects whose length can be recorded in the given level of detail, however, not their width.
- *Areas* represent objects whose planar dimensions may be recorded and visualized in the given level of detail.
- *Surface* illustrates the shape of an object similarly to area, however, it enables to record also changes in height, or the whole height, potentially also another parameter that continuously changes in the given space.

Geographic objects and phenomena in representative models are presented through sets of data layers. Data layers—from the point of data format—may be vector or raster, from the point of space dimensions they may be two-dimensional (2D) or three-dimensional (3D). Special data layers contain image data received from sensors placed usually on planes, satellite, or pilotless aircrafts. These data are of great importance for decision-making processes and they are very common in them (Štampach, Kubíček, & Herman, 2015). With regards to the focus and extent of this publication, however, they will not be considered.

2.1.2 Vector Models

Objects in the vector format are expressed as:

– Individual points,
– Lines objects,
– Areas (polygons).

Basic information unit is a *individual point* whose position in the given geodetic reference positioning, as well as the elevation system and the given projection is fixed. If from the point of view of logical causality there are more individual points belonging together, it is possible to define also an object created from these individual points as multipoint.

A *line object* is created by a set of points that characterize its geometry. Each line object begins in the initial point and ends in the final point. These points are called nodes. The line itself is called an edge. With the help of nodes, there are higher hierarchical structures created, usually by introducing topological characteristics.

Polygons are created by at least one line (edge line) with the same initial and end points. However, there even may be more edge lines, but they must always create closed areas.

Thematic characteristics of objects are saved in tables of characteristics (attributes). Number of attributes is basically unlimited and for each object it is possible to save all necessary thematic data.

Topological characteristics of vector data are their important feature. Topological characteristics enable to model connections between objects and make simple or complex spatial analyses, including solving traffic tasks, such as the shortest or fastest route. Vector data in GIS may have various topological levels that come from the theory of graphs:

– level 0—Spaghetti,
– level 1—Chain Node,
– level 2—Planar Graph,
– level 3—Full Topology.

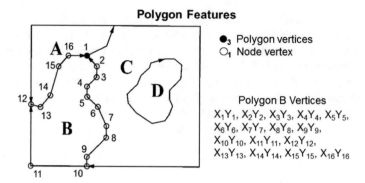

Fig. 2.1 Topological elements and relationships (ArcGIS, 2015)

Different topological levels of vector data enable to choose such an expression of topological relationships that corresponds to complexity, completeness and consistency of geographic data.

Topological level 0 (Spaghetti) does not describe any topological relationships between objects. *Topological level 1* (Chain Nodes) includes a topological relationship between nodes and edges when each edge must begin and end with a node. Edges may cross each other at any place. This level is useful for the description and analysis of connections. Another *topological level 2* (Planar Graph) sets a rule that edges may cross only in nodes. It enables to derive and work with a relationship of neighborhood, despite not being directly included in this data structure. The highest one is *topological level 3* (Full Topology). Here there is a topological entity "Wall" included. In this level it is possible to describe also topological relationships wall—wall as well as wall—node. This way, relationships of neighborhood and topological characteristics "Left- (right-) side wall" and "Node included in the wall" are included. The picture (Fig. 2.1) shows an example of topological elements and relationships; the picture (Fig. 2.2) illustrates a record of topological relationships in the GIS database.

An example of a vector model is in the picture (Fig. 2.3).

Vector format is usually used for modelling of discreet objects Kresse and Danko (2012), ArcGIS (2015).

2.1.3 Raster Models

Raster models are based on files in raster format. Basic unit here is a *raster cell—pixel*. Pixels are organized in rows and columns and each pixel carries information, e.g. about elevation, air quality, surface characteristics, etc. From the point of view of geographic objects and phenomena that are represented by raster models, this information may be discreet or continuous. Raster files may be expressed as matrix with a certain number of rows and columns. If they are worked with in the

Topological Elements and Relationships

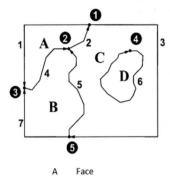

Face	Edges	Nodes
A	1,2,4	1,2,3
B	4,5,7	2,5,3
C	2,3,5,0,6	1,5,2,0,4
D	6	4

Edge	Left Face	Right Face	From-Node	To-Node
1	A	-	1	3
2	A	C	2	1
3	-	C	1	5
4	A	B	3	2
5	B	C	5	2
6	C	D	4	4
7	-	B	5	3

A   Face
1   Edge
3   Node
↗   Direction of edge

Fig. 2.2   Record of topological relationships in GIS database (ArcGIS, 2015)

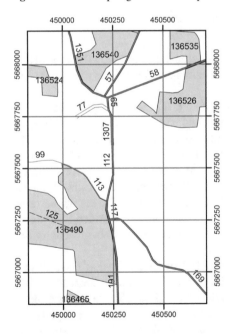

Transport Network			
Obj_ID	Obj_Code	SMC	WD1
58	AP030	018	5

Vegetation			
Obj_ID	Obj_Code	LES	HGT
136490	EC015	201	18

The properties of objects are stored in a database table and codded according to given codding catalogue. The catalogue of topographic objects (MoD-GeoS, KTO DMU25, 2013) was used for this example. The Obj_ID ensures a link between geometry and database table. In the example the object code AP030 is Road, EC015 is Forest. SMC is Surface (018 – paving), WD1 is Width in meters, LES is Stage of forest (201 – grown) and HGT is Hight in meters.
Position is given by coordinates (E, N in WGS84/UTM).

Fig. 2.3   Vector model and table of characteristics

environment of geo-information systems, these files are transformed into a suitable geodetic reference system and alternatively also into a suitable projection. Raster pixels then have also positional information.

In raster models of discreet objects (e.g. forest vegetation, hydrologic network, communication system, soil types, etc.) there is information in the individual raster pixels in the form of thematic data and in neighboring raster pixels they do not have to follow each other thematically.

With continuous models of natural or social phenomena, such as temperature, pressure, emission, noise, population density, etc., raster pixels contain information about the value of the given phenomenon in the given pixel. Values of neighboring pixels then follow each other and with the help of appropriate procedures it is possible to retrospectively interpret the process and arrangement of the modelled phenomenon.

From the point of view of spatial analyses, raster models have significant advantages:

- Simple data structure models topological characteristics of objects in a simpler way than vector models.
- Simple data structure enables the use of robust statistical and spatial analyses.
- Relatively easy representation of continuous fields and surfaces enables to analyze these surfaces and detect their characteristics.
- Ability to unify saving of points, lines, areas, and surfaces.
- Ability to quickly create overlays of various raster fields.

The disadvantage of raster data is their limited accuracy given by the level of data detail and pixels size and relatively high requirements of the memory for saving and manipulation of these files.

As already said, the basic unit of a raster field is a raster cell—pixel that carries a value. Raster cell may be of various shape (square, rectangle, triangle, hexagon, octagon), however, most often a square with a defined side is used (e.g. 1 × 1 m, 100 × 100 m, 1 × 1 km etc.). The value of a raster cell may be an integer constant or a real number. So, it may get positive as well as negative values. According to the character of the object or phenomena that it represents, this value is related either towards the centre of the raster cell or it is considered for the whole raster cell. Integer value usually represents objects or phenomena classified into pre-set categories, real number expresses a measured or interpolated value of continuous object or phenomenon (Fig. 2.4).

The dimension of raster cells significantly influences the *level of detail* (LOD) of the modelled object or phenomenon. The smaller the physical dimension of the raster cell, the smoother and more detailed the model (Fig. 2.5). On the other hand, with a decreasing dimension of a raster cell, their number increases and thus also requirements on manipulation with the whole raster field are higher which manifests in higher demands on the performance of computer systems and transmission lines, or it manifests in deceleration of data processing.

To accelerate work with raster files while keeping their level of detail, basically two methods are used. One of them is based on a division of the whole raster file into smaller parts (so-called tiling), manipulation with which is easier and faster. This method maintains the LOD but works with smaller data files. The other method is based on an idea that when working with the whole file it is not necessary to use

Fig. 2.4 Possibilities relate values to raster cells: left—discreet, right—continuous (ArcGIS, 2015)

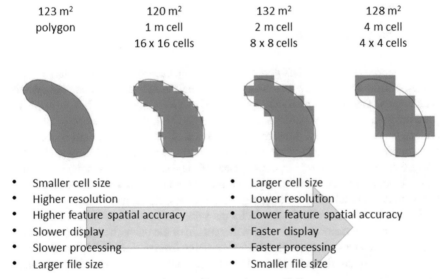

123 m²	120 m²	132 m²	128 m²
polygon	1 m cell	2 m cell	4 m cell
	16 x 16 cells	8 x 8 cells	4 x 4 cells

- Smaller cell size
- Higher resolution
- Higher feature spatial accuracy
- Slower display
- Slower processing
- Larger file size

- Larger cell size
- Lower resolution
- Lower feature spatial accuracy
- Faster display
- Faster processing
- Smaller file size

Comparing small versus large cell sizes

Fig. 2.5 Dependence of raster cell size and detail of the modelled object (ArcGIS, 2015)

maximal LOD for some tasks. Maximal LOD is needed when examining details, or when observing modelled objects or phenomena. This approach is called creation of pyramids (Fig. 2.6).

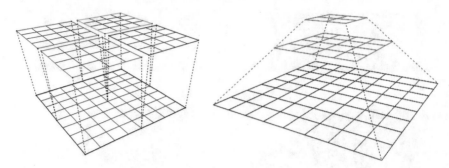

Fig. 2.6 Division of raster files: left—into tiles; right—division to files with smaller LOD—pyramids (Kresse & Danko, 2012)

Raster models are quite often used in spatial analyses, especially in multi-criterial analyses (ArcGIS, 2015; Kresse & Danko, 2012; Longley, Goodchild, Maguire, & Rhind, 2016).

2.1.4   Surface Models

Surface models are often considered to be a special category of representative models, especially because they are three-dimensional models of continuous objects or phenomena. The surface may model e.g. relief of a terrain, course of an atmospheric pressure field, atmospheric temperature in a certain altitude, etc. Surfaces are usually created by spatial interpolation from discreet points, on which required characteristics are measured (altitude, atmospheric temperature, level of its pollution, level of noise, etc.). From the point of view of shape of planes, it is possible to consider fully smooth models that do not have any continuities, and partially smooth models where there are discontinuities (barriers), e.g. terrain edges, risers, rocks, etc.

Surface models may be expressed by the same methods as raster models. In this case, the value of raster cell is the third dimension of the phenomenon in the position of this cell. However, also specific methods coming from vector models are often used for them. These are first of all models using contour lines and *triangular irregular networks* (TIN).

Contour lines are lines that connect places with the same value of the given phenomenon, e.g. elevation (isohypse), temperatures (isotherms), pressure (isobars), magnetic field of the Earth (isogons), etc. Contour lines are interpolated from a raster model or TIN model usually in a pre-set interval of the third dimension and they are saved as vector data where the attribute line is just the value of the given phenomenon (Fig. 2.7). These models are mostly used for smooth surfaces.

TIN model is based on a set of points with values of planar coordinates and height (or value of a modelled phenomenon in the given point) and a series of edges connecting these points into triangles. Thus the surface is modelled by an irregular

Fig. 2.7 Contour lines—example

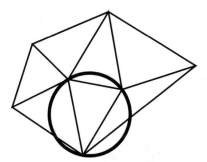

Fig. 2.8 Principle of TIN creation

triangle network. These created triangles must comply with so-called Delaunay criterion which requires that circumscribed circle of three vertexes (nodes) does not contain any other point. It means that each point is connected to its closest neighbor (Fig. 2.8).

TIN consists of nodes, edges, triangles, clipping polygons, and topology. Nodes are basic building blocks of TIN. Each node (Fig. 2.9) must have its height and it is connected to the closest neighbor by an edge (Fig. 2.10). As there are nodes with height at the ends of an edge, it is possible to calculate e.g. gradient along the edge (gradient of a phenomenon).

Series of edges create triangles. Each triangle describes behavior of a part of the surface and can be used for determination of gradient, view, area, and length. Taking triangles as a whole into consideration, it is possible to derive morphometric characteristics, e.g. profiles, visibilities, or hidden spaces.

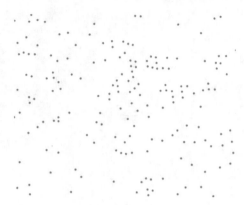

Fig. 2.9 Nodes

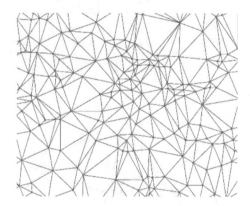

Fig. 2.10 Edges and triangles

Fig. 2.11 Clipping polygon

The envelope is formed by one or more polygons containing data for model construction. Such polygon defines the zone where interpolation process takes place. Outside the polygon, there is no interpolation (Fig. 2.11).

Triangles describe behavior of a part of surface and they may be used for the determination of gradient, view, surface, and length. Coordinates may be used for

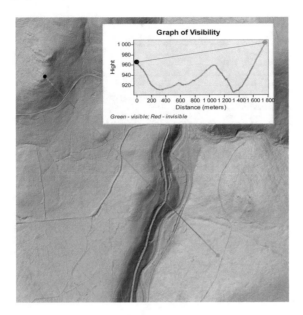

Fig. 2.12 Determination of line of sight and a profile graph. The green parts of the line mark visible sections, red ones—invisible sections

determination of information about the area of the given triangle, such as gradient, view, area of the surface, and length on surface. With consideration of triangles as a whole, it is possible to derive profiles, visibilities, or terrain views (Fig. 2.12).

TIN may be created from regularly or irregularly positioned points. If a surface with discontinuities is modelled, it is necessary to input these discontinuities as lines or areas, on which the calculation of interpolated triangles is interrupted. Discontinuities may be soft, hard, or they may be considered as breaks.

Soft discontinuities are used to maintain a known value of height along line objects and polygons and they form the edges of TIN. *Hard discontinuities* define interruptions of smoothness. They are used for the definition of water courses, ridges, coastlines, etc. For instance, they may define a sharp transition between the surface of a lake and a range. *Breaks* represent interruption of a surface continuity. They have more values of *height* in the same place on the top and the foot of the break. Breaks are not only vertical, but also horizontal. When saving, it is possible to present them as two parallel lines close to each other, containing values of height on top and at the foot.

To create the height model, special programs that are often a part of program systems for GIS are usually used.

An advantage of a TIN relief model is its variability in terms of resolution. In places where there is a small change in the values of height, a bigger number of points is chosen, and thus it is possible to describe also rather complicated terrain shapes. That is the reason why TIN is used as one of the basic methods for modelling the relief of terrain.

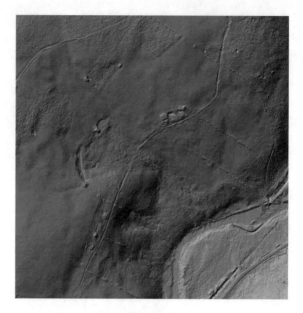

Fig. 2.13 Data representation of surface in TIN model

Apart from vector model also *raster model* may be used for modelling surfaces, especially in cases of analytical processes. Raster model may be created directly from the set of regularly or irregularly positioned points that in their characteristics carry the value of height or another parameter, from which the surface itself is interpolated. However, it is possible to use an already created TIN surface model and transform it into a raster model. The result is a continuous raster layer with pixels carrying information about the height.

With regards to the fundamental difference in data representation of a TIN model and a raster model, with raster model there is a degradation of information about the position, height, and character of the modelled surface, even also with a rather high level of detail of a raster model. This fact can be seen in the pictures (Figs. 2.13 and 2.14) that show the identical part of the surface of terrain relief.

Different data representation may manifest also in the results of all analytical procedures performed with these data. As an example, you can see an analysis of visibility between two points solved with a TIN model and with a raster model (Figs. 2.15 and 2.16).

Fig. 2.14 Data representation of surface in raster model with raster cell size of 1×1 m

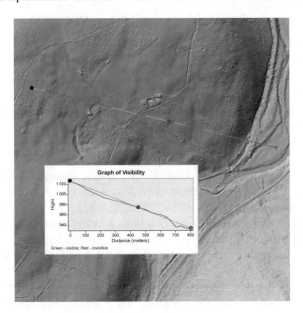

Fig. 2.15 Analysis of visibility with a TIN surface model

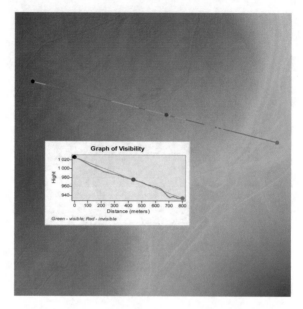

Fig. 2.16 Analysis of visibility with a surface raster model with pixel size 1 × 1 m

2.2 Assessment of Quality of Geospatial Data

The user of the source as well as newly created data should always get—apart from the spatial information itself—also information about its quality. With assessment of quality, it is possible to follow general scheme of quality components which assesses production-technological aspects that manifest in technical functionality, as well as operational and security aspects; in relation to a specific use of a product or service also reliability aspects (Fig. 2.17).

When assessing spatial data and information, it is necessary to modify the stated understanding of quality with regards to positional determination of the modelled objects and phenomena. Most systems of assessment of spatial data and information assess technical parameter of data or technological influence during their collection. However, it is good to apply also user's aspects of data and information, which assess mainly their utility value that is given by the quality of information and reliability when it is used.

In general, utility value or functionality of given product can be understood as the ability of the product to meet a user's need. Functionality consists of the functional effect of all the product properties that result from its nature as well as its phenomenal aspects (Miles, 1989).

As technical parameters, it is possible to assess the accuracy of positional information, often given by a mean positional error, or a mean error in the individual axis of coordinates. Assessment of thematic information depends on its character, i.e. if it is qualitative or quantitative data. Assessment of user's aspects follows the requirements set by the user and the degree of their fulfillment. The following paragraphs deal with possible ways how to assess the given requirements.

Fig. 2.17 General components of quality assessment

2.2.1 Concept of Quality of Geospatial Data According to International Standards

In general, in order to formulate the problem of assessment of quality of spatial data and the resulting geospatial information it is necessary to follow the recommendations of international organizations, such as ISO (International Organization for Standardization), OGC a DGIWG (ISO, 2006; Open GIS Consortium (OGC), 2010; DGIWG-103, 2008; DGIWG-500, 2010), that in the long term focus on the development of geoinformatics, and recommendations of the INSPIRE (Infrastructure for Spatial Information in Europe) directive (INSPIRE, 2011). Systems for quality assessment are being developed in these organization and consortiums. The problem of quality and standardization of DGI is dealt with also in Join Research Centres (JRC), which regularly publish their approach, e.g. in JRC (2010, 2011) or Reznik (2013). According to Handbook for implementation of ISO 19100 standards for quality assessment of geographic information, published by ISO/Technical committee 211 Geographic Information/Geomatics (ISO/TC2011), see ISO (2017), it is necessary to assess quality as a complex problem—of production—concerning technology and data source—as well as economical, that deals with the fulfillment of user's requirements in relation to economical possibility of the producer and success of the product on the market, and last but not least—legislative that deals with the acceptance of legal conditions for the use of DGI on national as well as international level, including respecting national and international standards (Fig. 2.18).

Technical functionality of spatial data is influenced especially by technological processes of preparation, collection and processing of spatial data that are done by

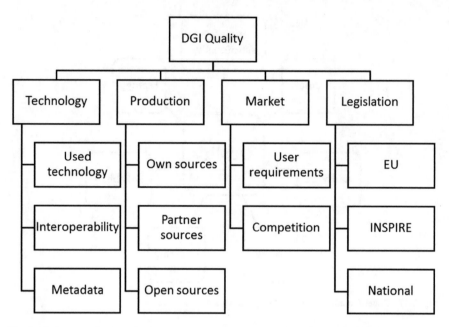

Fig. 2.18 Reasons for introducing standards for assessment of quality of geospatial information. Adapted according to Jacobsson & Giversen (2007)

the used systems, format of the saved data, etc. Data saved in one database do not have to be gathered within one organization, and even by one technology. On the contrary, within national and international co-operations data are often created by being defined according to one concept, however, in reality this concept might in production organizations be modified in a way so that on one hand it better meets the standards, but on the other it respects technical and technological conditions of the production organization. This fact is apparent in international projects (e.g. in border areas, regional or global type of INSPIRE, Vector Smart Map, Multinational Geospatial Co-Production Program etc.), in which it is possible to meet various approaches to definition of objects and their subsequent modelling. An example might be different perception of the importance of communication in conditions of Central Europe and Afghanistan.

Within ISO standards for DGI a system of evaluation of data quality was designed, it follows the recommendations of ISO/TC 211 (ISO, 2017). The base of evaluation of data quality is the recommendation set by norm ISO 19157 (ISO, 2013), which observes the following elements of quality.

During the assessment of the individual elements, however, it is necessary to specify methods of this assessment so that they are objective as well as they reflect also user aspects of data. One possible approach to assessment of data quality will be described in paragraph 4.2. Before that, however, it is necessary to mention the relationship of data quality in relation to its reliability which is manifested also in vagueness of used data and their attributes.

2.2.2 Modelling of Vagueness

Within sub-elements of data quality, accuracy of data presentations of objects in databases are evaluated. These accuracies are usually assessed by statistical methods when differences in characteristics of data representations of objects and their real characteristics in reality are determined. From these differences then random and systematic mistakes are discovered and the whole model is characterized for example by a mean error in coordinates, mean error in position, etc. (Kratochvíl, 2000). However, the vagueness of data representations of objects is manifested in position as well as in thematic characteristics of the concrete objects. That is the reason why it is good to consider this vagueness when working with concrete objects and to have corresponding models of vagueness ready (Hubacek, Kovarik, & Kratochvil, 2016).

As stated before, data representation of geographic object within GIS is a spatial element that is characterized by its geometry, corresponding attributes and relation with other surrounding objects. Borders of geographic objects, however, are only a consequence of human perception, not the real state of things. Even in cases when there is a real discreet border (ground plan of a building, runway, etc.), the border line may be inaccurate due to data vagueness or their interpretation. A significant part of objects is characterized by non-existence of a border, or this border is vague or variable—e.g. a border of forests, coastline of water areas, or a border of an area hit by emissions.

A similar situation takes place with data representation of object characteristics. They can be expressed for example by numbers expressing their physical characteristics in given units (height, depth, speed, bearing capacity, etc.) or they may be put into priorly set categories (height from – to, depth from – to, etc.). Real characteristics in the given time cannot be found from data representations because they may change in time.

Objects in spatial databases are usually represented without considering the mentioned vagueness. Their borders (geometric as well as thematic) are taken as crisp. While in real conditions, especially in spatial analyses, we should take into consideration the vagueness and inaccuracy of saved information in representative models as this vagueness significantly influences the reliability of analyses and consequently also decisions based on geospatial data support. Brown & Hauvelink (2007) suggest 2 types of inaccuracy that have a special significance in GIS—*spatial vagueness* and *attribute ambiguity*. In the first case it is the inability to find a precise placing of an object with a given theme, in the other it is the inability to confirm an accuracy of the given theme in the specific place. However, spatial and thematic data should be assessed independently. One of the options how to solve such problems is to use the theory of "*soft classification*", where also "*fuzzy approaches*" belong to.

2.2.2.1 Modelling of Spatial Vagueness

Spatial vagueness can be determined based on the difference between geometry saved in the data representation of an object and the real geometry in the real world or geometry in data representation of an object in another database that is at least by one order more accurate than in the evaluated representation.

Spatial vagueness in case of one-point objects leads to a unit shift in the position of an object in several possible ways as a result of coordinates used to its description (x, y, z). Spatial vagueness of solid objects on the other hand represent rotation around the original position, as well as simple shifts. With unsolid objects, vagueness of individual definition points leads to complex changes of shape as well as position. It happens especially when position vagueness of individual points is partially or completely independent.

One of the possible approaches of solution of positional vagueness in geographic objects is the use of the theory of probability and estimation with the help of continuous *probability distribution function* (*PDF*) for all basic vertices (points) of an object. For two-dimensional point object, *PDF* of coordinates x and y is then defined as:

$$PDF_{XY}(x, y) = Prob(X \leq x, Y \leq y),$$

where X and Y are random variables representing a real, however unknown, position of the given point (Fig. 2.19). Average and standard deviation of coordinates x and y are important parameters of *PDF*. While the average determines the position of a point in all directions, standard deviation determines its vagueness.

For solid objects consisting of more connected points it is possible to define a shift and rotation of an object around one reference point. Left part in Fig. 2.20 gives an example of 4 random projections of a position of a vague solid object whose vagueness in direction x is higher than in direction y and for which rotation vagueness is relatively small.

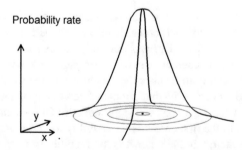

Fig. 2.19 Schematic demonstration of probability distribution of positional vagueness. Adapted according to Shekar & Xiong (2008)

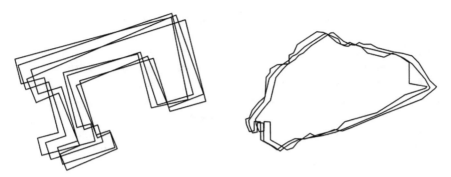

Fig. 2.20 Demonstration of modelling of positional vagueness for vague solid objects (left) and for vague changing objects (right)

For changing objects, it is necessary to define a continuous function of division of probabilities for all basic points which create the object (Kubíček et al., 2017; Maturo, Fortuna, & Di Battista, 2018). Neighboring points will hardly ever change independently because in their determination they are often similar in neighboring positions. This implies the need to model relationships or correlation between vague points. An example of a correlation model dependent on the distance of a point can be a semivariogram. Right part in Fig. 2.20 shows an example of a presumed strong spatial relation (correlation) between neighboring points. The correlation is not complete as 4 performed realizations of the object are deformed.

In both cases it is rather complicated to express a definite parameter that would describe the positional vagueness of the border similarly to *PDF* of a point object.

2.2.2.2 Modelling of Attribute Vagueness

Vagueness of attributes is defined as closeness of the attribute value to the real value in the real world where individual objects are usually of complex nature and where for example material creating an object is spatially heterogeneous and borders of an object are not usually sharply set. The core and level of attribute vagueness is influenced by many factors:

– Attributes of geospatial object,
– Way of determination and identification of attributes,
– Method and technology used for obtaining or measuring of attribute value,
– Mathematical core of spatial analyses and model procedures used for attribute data.

Attribute vagueness may consequently influence the quality of geo-spatial decision-making. In many GIS applications the theoretic requirements on accuracy of attributes are even higher than requirements on positional accuracy.

2.2.2.3 Modelling of Vagueness with the Help of Fuzzy Sets

One of the ways how to involve positional as well as attribute vagueness into solutions of spatial analyses to support decision-making processes is the use of fuzzy sets.

Fuzzy sets (vague sets) offer a framework for processing predicates whose level of truth is given in degrees ("True to certain degree") and uncertainty is expressed also in degrees (Davvaz & Cristea, 2015; Hoskova & Cristea, 2010). The concept of fuzzy sets deals with representation of classes whose borders are not clearly (sharply) set. With absence of a sharp border dividing the set from the surroundings there is a problem of unambiguous determination of affiliation of an element in a set and in its complement.

Not sharp or fuzzy sets are then sets, or classes, that do not have sharp boundaries. With spatial data it means that at considered places the transition between affiliation and non-affiliation in a set may be gradual. Fuzzy set may be characterized by fuzzy levels of affiliation in interval <0, 1> which express a gradual growth of affiliation from non-affiliate up to total affiliation. They can be defined using an affiliation function (membership function).

Fuzzy sets can be used for solution of positional vagueness, as well as for solution of vagueness of characteristics. The following simple example shows differences in the approach to data representations and their vagueness. Vector data DMU25 are produced by Geographic service of Czech Army whose data representation is given in MoD-GeoS (2013), image data are provided as web service run by State Administration of Land Surveying and Cadastre of the Czech Republic (CUZK, 2018). The presented area comes from Military training area Libavá in north-east part of the Czech Republic and it is in scale 1:1,000. Program system ArcGIS 10.2. (Esri, 2013) was chosen for illustration.

Picture (Fig. 2.21) is a part of countryside where two main objects are an unpaved road and a forest.

Their data representation is shown in picture (Fig. 2.22), from which it is obvious that neither of the objects is in the same position as in reality.

Considering the classification of both objects as a accuracy class, where the medium positional error is 5 m, it is possible to state that the real appearance of the road and forest border with 95% probability will be in a belt up to 10 m right and left from the data representation of the object or its border. This fact can be expressed by a buffer 20 m wide whose axis is the data representation, in this case object "road" (Fig. 2.23).

The disadvantage of such solution is that all points inside of the buffer carry the same probability of affiliation to object "road".

However, it is obvious that affiliation of any point inside of the buffer should be lower with growing distance from the data representation of the road. For such understanding of affiliation fuzzy sets can be used. The result of expression of such understanding of vagueness with the help of fuzzy sets is obvious from the picture (Fig. 2.24).

The affiliation of the surroundings of data representation of the road to object "road" in the picture (Fig. 2.24) is displayed as decreasing intensity of black color.

Fig. 2.21   Countryside in Military training area Libavá displayed in image data (CUZK, 2018)

Fig. 2.22   Countryside and its data representation displayed over image data (red dashed line—road, green—forest, green dots—scrubs

Fig. 2.23 Buffer of road in distance of 10 m to both sides (yellow lines and dots)

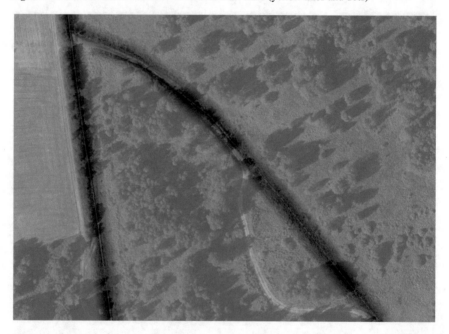

Fig. 2.24 Affiliation of the surroundings of data representation of road to the object "road"

Maximal affiliation is in the axis of the road and it decreases within 10 m to both sides. For the example, a linear function of affiliation was chosen (ArcGIS, 2015). From the picture it is obvious that none of the used methods removes errors in data representation given by the wrongs digitalization (see left top corner of the picture).

Similarly, it is possible to work with vagueness with all objects.

2.2.3 *Evaluation of Data Quality*

Evaluation of data quality follows the recommendation given by ISO 19157 (ISO, 2013) which observes elements of quality stated in the table (Table 2.1).

Produced data are meant especially for consumption, i.e. their use in concrete procedures of planning and control, spatial analyses, etc. From the point of view of the user, also systems of evaluation of used data given especially by evaluation of their utility value are very important.

Technical functionality can be evaluated in general without the need to know the concrete task. However, it is necessary to evaluate procedures or usage of spatial information, other components of quality, especially reliability, with regards to concrete usage in a concrete process. The quality of spatial data and information may be evaluated by the criteria that follow ISO 19157 (see Table 2.2), and which were complemented with specifics bound to evaluation of user's point of view.

Table 2.1 Elements of quality according to norm ISO 19157

Element of data quality	Sub-element of data quality
Completeness	Adding
	Omitting
Logical consistency	Conceptual consistency
	Domain consistency
	Data format consistency
	Topological consistency
Positional accuracy	Absolute accuracy
	Relative accuracy
	Raster data accuracy
Time accuracy	Accuracy of time measurement
	Time consistency
	Obsolescence of data
Attribute accuracy	Correctness and accuracy of classification
	Correctness of classification of attributes that cannot be expressed quantitatively
	Accuracy of quantitative attributes
Data administration	

Table 2.2 Criteria of spatial data quality and information

Group of main characteristics—main criteria	Characteristics of partial criteria	Definition	Quality criterion
Content of database—k_1	Completeness of real world model—k_{11}	Conformity of the defined model with users' requirements	Percentage rate of incomplete information
	Compliance of required resolution for geometric and attribute data—k_{12}	Conformity with the required level of detail of data	Percentage rate of objects with not fulfilled conditions of geometric level of detail—k_{121}
			Percentage rate of objects with not fulfilled conditions of thematic level of detail—k_{122}
Technical quality of database content—k_2	Transparency of data sources and methods for secondary data derivation—k_{21}	Transparency of source materials on primary data collection	Level of availability of information about used sources—k_{211}
		Transparency of methods and models used for secondary data derivation	Level of availability of information about used methods—k_{212}
	Positional accuracy —k_{22}	Compliance with declared horizontal accuracy	Percentage of objects with satisfactory horizontal accuracy —k_{221}
		Compliance with declared elevation accuracy	Percentage of objects with satisfactory elevation accuracy—k_{222}
	Attribute accuracy—k_{23}	Compliance with declared accuracy of attribute data	Percentage of objects with satisfactory attribute accuracy

(continued)

Table 2.2 (continued)

Group of main characteristics—main criteria	Characteristics of partial criteria	Definition	Quality criterion
	Logical consistency of database—k_{24}	Degree of adherence of geographic data (data structure, features, attributes, and relationships) to the models and schemas (conceptual model, conceptual schema, application schema, and data model)	Percentage of objects with topological consistency—k_{241}
			Percentage of objects with attribute consistency—k_{242}
			Percentage of objects with time consistency—k_{243}
	Data completeness—k_{25}	Degree of adherence of the entirety of geographic data (features, attributes, and relationships) to the entirety of the modelled part of landscape	Percentage of missing objects of surplus objects—k_{251}
			Percentage of properties of thematic objects that are complete—k_{252}
Database timeliness—k_3	Degree of obsolescence of geospatial data with regards to changes in the given environment—k_3	Date of last update	Number of changes
			Time since last change
			…

(continued)

Table 2.2 (continued)

Group of main characteristics—main criteria	Characteristics of partial criteria	Definition	Quality criterion
Area importance —k_4	Value of inverse distance of evaluated area and objects of interest—k_{4i}	Significance of area from the point of view of fulfilled functions	Geographic position of evaluated part of area
			Entry corridors into the interested part of area
			Number and character of obstacles
			Centers of industrial production
			Population density
			Distribution of systems and equipment securing the defense of area
			…
User friendliness —k_5	Data standardization—k_{51}	Compliance with declared standards	Percentage rate of non-compliant standards
	Data independence on program environment—k_{52}	Level of independence on types of application programs	
	Data security against damage or abuse—k_{53}	Used degree of data security and its level	Degree of protection of access rights—k_{531}
			Degree of protection of copyright—k_{532}
			Degree of physical protection of data—k_{533}

References

ArcGIS. (2015). ArcGISHelp. Retrieved from www.esri.com.

Brown, J. D., & Hauvelink, G. M. (2007). The Data Uncertainty Engine (DUE): A software tool for assessing and simulating environmental variables. *Computers & Geosciences, 33*(2), 172–190.

CUZK. (2018). (State Administration of Land Surveying and Cadastre, Land Survey Office). Retrieved 10 2017, from Geoportal CUZK—Access to map products and services: http://geo portal.cuzk.cz/.

Davvaz, B., & Cristea, I. (2015). Fuzzy algebraic hyperstructures. *Studies in Fuzziness and soft computing, 321.*

DGIWG-103. (2008). *Digital Geographic Information Exchange Standard (DIGEST)—Metadata Profile of ISO19115 and ISO 19139* (2.0.0—16 December 2008 ed.). DGIWG. Retrieved from https://www.dgiwg.org/dgiwg/.

DGIWG-500. (2010). *Implementation guide to the DGIWG Feature Data Dictionary (DFDD)* (2.2.2—19 July 2010 ed.). DGIWG.

Esri. (2013). ArcGIS user documentation. Copyright © 1995–2013 Esri.

Fotheringham, A. S. (2000). *Spatial models and GIS*. London: Taylor & Francis.

Hoskova, S., & Cristea, I. (2010). Fuzzy pseudotopological hypergroupoids. *Iranian Journal of fuzzy sets, 4*(6), 11–19.

Hubacek, M., Kovarik, V., & Kratochvil, V. (2016). Analysis of influence of terrain relief roughness on DEM accuracy generated from LIDAR in the Czech Republic territory. In *International Archives of the Photogrammetry, Remote Sensing & Spatial Information Sciences, XLI-B4* (Vol. 41, pp. 25–30). Prague: ISPRS. https://doi.org/10.5194/isprsarchives-xli-b4-25-2016.

ICA. (2014). *International Cartographic Association*. Retrieved July 2014, from International Cartographic Association: http://icaci.org/mission/.

INSPIRE. (2011). *European Commission—INSPIRE*. (European Commission) Retrieved 12 23, 2011, from European Commission—INSPIRE: http://inspire.jrc.ec.europa.eu/index.cfm.

ISO. (2006). *ISO 19138—geographic information—data quality measures*. Retrieved 2011, from ISO—Interational Organization for Standardization: http://www.iso.org/iso/catalogue_detail.htm?csnumber=32556.

ISO. (2013). *ISO 19157:2013 geographic information—data quality*. Retrieved from International Organization for Standardization: https://www.iso.org/standard/32575.html.

ISO. (2017). *ISO/TC 211 geographic information/geomatics*. Retrieved from International Organization for Standardization: https://www.iso.org/committee/54904/x/catalogue/.

Jacobsson, A., & Giversen, J. (2007). *Eurogeographics*. Retrieved 2009, from http://www.eurogeographics.org/documents/Guidelines_ISO19100_Quality.pdf.

JRC. (2010). *INSPIRE data specification on hydrography—guidelines*. Retrieved 11 2011, from EU—INSPIRE–INSPIRE Registry: http://inspire.jrc.ec.europa.eu/documents/Data_Specifications/INSPIRE_DataSpecification_HY_v3.0.1.pdf.

JRC. (2011). *Data specification on natural risk zones—draft guidelines*. Retrieved 2012, from EU—INSPIRE–INSPIRE Registry: http://inspire.jrc.ec.europa.eu/documents/Data_Specifications/.

Kratochvíl, V. (2000). *Geodetic networks—application of the least squares method and the transformation of coordinates (in Czech)*. Brno: Military Academy.

Kresse, W., & Danko, D. M. (2012). *Handbook of geographic information*. Berlin Heidelberg: Springer. https://doi.org/10.1007/978-3-540-72680-7..

Kubíček, P., Šašinka, Č., Stachoň, Z., Štěrba, Z., Apeltauer, J., & Urbánek, T. (2017). Cartographic design and usability of visual variables for linear features. *Cartographic Journal, 54*(1), 91–102. https://doi.org/10.1080/00087041.2016.1168141.

Lloyd, C. D. (2011). *Local models for spatial analysis*. Boca Raton: Taylor & Francis LLC.

Longley, P. A., Goodchild, M. J., Maguire, D. J., & Rhind, D. W. (2016). *Geographic Information Science and Systems* (4th ed.). Wiley.

Maturo, F., Fortuna, F., & Di Battista, T. (2018). Testing equality of functions across multiple experimental conditions for different ability. *Social Indicators Research*, 1–21.

Miles, L. D. (1989). *Techniques of value analysis engineering* (3rd ed.). USA: Eleanor Miles Walker.

MoD-GeoS. (2013). *Catalogue of the topographic objects DMU25* (7.3 ed.). Dobruska: Ministry of Defence of the Czech Republic, Geographic Service.

Open GIS Consortium (OGC). (2010, October 18). *WEB processing service*. Retrieved 09 18, 2015, from OGC: Making location count: http://www.opengeospatial.org/standards/wps.

Reznik, T. (2013). Geographic information in the age of the INSPIRE directive: Discovery, download and use for geographical research. *Geografie, 118*(1), 77–93.

Shekar, S., & Xiong, H. (2008). *Encyclopedia of GIS*. Berlin: Springer.

Sklenak, V., Berka, P., Rauch, J., Strossa, P., & Svatek, V. (2001). *Data, information, knowledge, and internet (data, informace, znalosti a Internet, In Czech)*. Praha: C.H. Beck.

Snyder, J. P. (1987). *Map projections—a working manual*. Washington, USA: U.S.Geological Survey.

Štampach, R., Kubíček, P., & Herman, L. (2015). Dynamic visualization of sensor measurements: Context based approach. *Quaestiones Geographicae, 34*(3), 117–128. https://doi.org/10.1515/quageo-2015-0020.

Talhofer, V. (2007). *Elements of map projections (Základy matematické kartografie, In Czech)*. Brno: Univerzita obrany.

Tomlin, C. D. (1990). *Geographic information systems and cartographic modeling*. New Jersey: Prentice-Hall Inc.

Chapter 3
Multi-criteria Analysis

For the overall assessment of the quality of DGI, it is possible to use the general view of the utility value of a product based on general principles of multi-criteria analysis. Its brief description is given in the following paragraphs.

Multi-criteria Decisional Analysis (MCA) deals with the evaluation of possible alternatives according to several criteria, whereas an alternative evaluated according to one criterion is not usually best rated according to another criterion.

Methods of multi-criteria decision-making then solve conflicts between mutually contradictory criteria. This is a method that aims to summarize and sort information about variant projects. The application of the multi-criteria research method will allow optimization of the comparison of the specified criteria.

Multi-criteria analysis is a benchmarking tool that takes different views into consideration. These views are addressed through factors that affect the overall state of the product or activity. Factors can be both measurable (dimensions, weight, speed, etc.), as well as unmeasurable or difficult to classify, which are difficult to appreciate, such as the state of the environment, customer satisfaction, etc.

This analysis is mainly used in the conception phase, especially in complex cases and especially when different criteria (e.g. new jobs versus the impact of their establishment on the environment) or when the choice between the criteria is complex. Generally, this analysis is used in ex-ante evaluations, less often used in ad hoc or ex-post evaluations.

Multi-criteria decision-making arises where the decision-maker evaluates the consequences of his/her choice according to several criteria, namely quantitative criteria, which are usually expressed on natural scales (we are talking about numerical criteria) or qualitative criteria. For qualitative criteria, we introduce a suitable scale, such as a scale of scale or scale: very high-high-average-low-very low and at the same time defining the direction of the better evaluation, i.e. whether the maximum or minimum value is better, i.e. decreasing or rising values.

The alternative denotes each solution from the selection set. The criterion is the property that we assess for the given alternative. Each criterion is assigned a weight that expresses the importance of individual criteria relative to others.

© Springer International Publishing AG, part of Springer Nature 2019

V. Talhofer et al., *Quality of Spatial Data in Command and Control System*, Studies in Systems, Decision and Control 168, https://doi.org/10.1007/978-3-319-94562-0_3

Table 3.1 Evaluation matrix

Criteria/variants	X	Y	Z	Final assessment
A				
B				
C				
D				

This method consists of these steps:

1. Alternatives are identified. E.g. it is a problem of unemployment in a naturally preserved area, and variants can include an amusement park, a new factory or a combination of agro-tourism and local craft support.
2. Criteria (factors) that will be decisive when making a decision are set. E.g. in this case, the number, quality, job stability, environmental impacts, benefits for the local economy, infrastructure costs.
3. Detailed assessment of the impact of each alternative on each criterion. Where possible, it will be expressed in figures (not necessarily money).
4. Each factor (s) is determined by its relative weight (materiality). Indeed, indicators of the significance of the main impacts will emerge. This significance (i.e. the importance of individual factors) can be determined, for example, by research of local residents.
5. Such processed alternatives are evaluated in a public discussion (Skulinova, 2013).
6. The initial step of each MCA analysis is to construct an evaluation matrix, the elements of which reflect the evaluation of the individual criteria for each alternative (Table 3.1). Elements of this matrix do not have to be numbers.

Models of multi-criterial decision-making project decision-making problems, in which consequences of the decision evaluate according to more criteria. Consideration of more criteria in evaluation brings difficulties into solution of a problem which stem from the general contradiction of criteria. If all criteria pointed to the same solution, it would be enough to have just one of them in order to make the most suitable decision. The purpose of the models in such situations is either to find "the best" option according to all considered points of view, exclusion of ineffective options or creation of ordered set of options.

Approaches to multi-criteria decision-making vary according to the nature of the set of variants or admissible solutions. Depending on how it is entered, two groups of these models can be distinguished:

1. Models of multi-criterial evaluation of variants are entered using the final list of variants and their evaluation by individual criteria.
2. Multi-criteria programming models have a set of variants with infinitely many elements expressed by limiting conditions, and the evaluation of each variant is given by the individual criteria functions.

By deciding in the theory of multi-criteria analysis of variants, we understand choosing one or more variants from a set of acceptable variants and recommending them to be implemented. The decision maker should proceed objectively in the selection of the variants, using the various methods and methods of analysis of variants.

Sometimes it is possible to separate the person from the person taskmaster of investigator (analyst). This process has its advantages and disadvantages. The advantage is that the analyst is rarely interested in the outcome of the decision, and therefore proceeds most objectively. The disadvantage may be the fact that the analyst is not familiar with all the details of a task that could not be captured by modelling. As a result, the recommendation may be the objectively "best" option, but virtually another alternative would be better, for example, the second best, especially with small differences in values of the aggregate decision criteria.

Typical MCA applications are selections of a supplier of some goods or services, personnel selections, or consumer decisions.

We use it in project management in various decision situations:

- In the study of opportunities, we need to choose from several suitable options or to avoid the worst threat.
- In a feasibility study, we need to choose from several implementation options of the project.
- When selecting the project, we choose the most appropriate way of realization of activities, the best supplier of purchasing components or services, etc. (Lacko, 2008).

MCA supports, among other things, application appendices to the MS Excel program that provide resources for multi-criteria analysis.

3.1 Multi-criteria Assurance Analysis

Let's first define the basic concepts that will be used in this chapter. *Decision* is understood as the choice of one or more variants from the set of all acceptable variants. The *decision maker* is the body responsible for making this decision. In the tasks of multi-criterial analysis of variants, the final (discrete) set of m variants is given, they are evaluated according to n criteria. The goal is to make a decision on which option is best ranked according to the criteria. This is the so-called *optimal variant*. Variants can be sorted from the best to the worst or divided into effective and inefficient variants.

Variants (alternatives) are therefore specific decision options that are feasible. We denote them A_i (for $i = 1, 2, \ldots, m$).

Criteria K_j (for $j = 1, 2, \ldots, n$) are the aspects by which variants are judged. If the evaluation of the variants according to the criteria is quantified, we sort the data into a so-called *critical matrix*, which is called e.g. $Y = (y_{ij})$. The elements of this matrix express the evaluation of the ith variant according to the jth criterion. Matrix rows match to variants, columns to criteria.

Criteria according to their nature are either maximizing or minimizing. Prior to the start of the evaluation, it is necessary to convert all criteria to one type, i.e., either all maximization or minimization criteria will be. For example, if we want to convert the minimization criterion to maximization, one of the options in this procedure is to select the largest number in the column of the appropriate criterion and subtract from that number the other criteria in the given column. The result is the linear distance of the actual value from the worst value, the greater the distance, the better (the criterion is maximizing). However, if we use one of a wide range of software products when processing, it is enough just to enter the type of individual criteria, and the program will standardize all the criteria on the yield type itself.

Another possible division of criteria is by their *quantifiability*. In this case, we distinguish criteria:

- *Quantitative*—objectively measurable data (e.g. monthly salary, travel time, start time),
- *Qualitative*—cannot be measured objectively, the variants are evaluated verbally, therefore it is necessary to use various scoring scales or relative evaluation of variants (e.g. possibility of further professional growth).

The importance of a particular criterion in comparison with other criteria is expressed by its *preference*. In this regard, we assess:

- *Aspiration level*—the value of the criterion to be achieved,
- *Order of criteria* (ordinal information on criteria)—sequence of criteria from the most important to the least important,
- *Criteria weights*—cardinal criteria information; the weight is the value of the interval $\langle 0, 1 \rangle$ and expresses the relative importance of the criterion compared to the others,
- *Compensation of the criterial values* expressed by the degree of substitution between the criterion values (it is possible to offset the bad criteria according to one criterion with better values according to another criterion).

Moreover, it is possible to divide individual variants according to their properties:

- *Dominant variant*—if all the criteria are maximizing, the variant A_i is dominated by variant A_j if there exists at least one criterion K_l, such that $y_{il} > y_{jl}$, while for other criteria it is true that $(y_{i1}, y_{j2}, \ldots, y_{in}) \geq (y_{j1}, y_{j2}, \ldots, y_{jn})$.
- *Pareto variant, non-dominant variant*—a variant that is not dominated by any other variant.
- *Ideal variant*—a hypothetical or real variant that achieves the best possible value in all criteria. Such a variant would dominate all other variants. There is no such real variant in the list of possible variants, therefore this variant is only hypothetical. If it really existed, the applicant would choose it and would not have to search for a compromise solution.

- *Basal variant*—a hypothetical or real variant, whose evaluation is the worst according to all criteria. Such a variant would be dominated by other variants. Such a variant is not a real variant, the applicant would otherwise be able to exclude it.
- *Compromise variant*—The only non-dominant option recommended for solution. Each compromise option should meet the following rules:

- Non-dominance—the variant must not be dominated by another variant,
- Invariance relative to the order of criteria—the order of criteria does not affect the selection of the compromise variant,
- Invariance with respect to the criterial values—if all elements are added to the same number (or multiplied by the same number), the set of selected variants or the selected variant must not be changed,
- Independence on identical values of the same criterion—if a criterion whose values are roughly the same for all variants, the set of selected variants must not be changed,
- Invariance with respect to the added dominated variants—if a dominated variant is added into a set of variants, the selected compromise variant must not be changed,
- Determination—according to each approach, at least one variant must be selected as a compromise,
- Unambiguousness—the chosen procedure gives a clear result, one option marks as compromise.

3.1.1 Methods of Determination of Criteria Weights

Most multi-criteria decision-making methods require that the criteria are distinguished from the point of view of their significance. One possibility is the numerical expression of this significance by means of so-called weights (the more significant the criterion is, the greater the weight). In order to the weight of criteria, determined by different methods, by various experts, were comparable, we express them in the normalized values w_j which can be calculated as follows:

$$w_j = \frac{v_j}{\sum_{k=1}^{n} v_k}, j = 1, 2, \ldots, n,$$

where

- v_j, k_j is the weight of criterion,
- n is the number of all considered criteria.

Normalized weights represent non-negative numbers whose sum equals one.

Methods for determining the weights of the criteria can be divided according to the information needed to determine the weights:

- The decision-maker cannot determine the preference—if the decision maker is unable to distinguish the importance of each criterion, the same weight is assigned to all criteria.

Table 3.2 Expression of preferences

Expression of preferences	
Numeric	Verbal
1	The criteria are equally important
3	The first criterion is slightly more important than the other
5	The first criterion is significantly more important than the other
7	The first criterion is much stronger than the other
9	The first criterion is absolutely more important than the other

- The decision maker has ordinance information about the criteria—in such a case the decision-maker is able to determine the order of importance of the criteria. Methods requiring ordinal information on criteria include the order method and the Fuller's method.
- The decision-maker has the cardinal information about the criteria—The decision maker knows not only the order but also the spacing in the order of preferences between the individual criteria. Methods based on this principle include the scoring method and Saaty's method.

For self-evaluation of the variants, it is enough for the decision-maker to choose one method. There are many methods for determination of the weights of the criteria, they are treated in detail in the literature, so we give here only their list, and a brief description of some of them.

1. Method of order
2. Fuller's method
3. Method of Quantitative Pair Comparison—Saaty's Method
4. Logarithmic least squares method
5. Successive weighting method

Satty's method—this method, in addition to the selection of a preferred criterion, determines the size of this preference for each pair of criteria (Saaty, 1980; Saaty & Vargas, 2000). In order to express the size of Saaty's preference, a point scale is recommended (Table 3.2).

For a more sensitive expression of preferences, it is possible to use intermediate levels: 2, 4, 6, 8.

Successive weighting method—with a large number of criteria, it is appropriate to group the criteria into subgroups according to the relationship of their material content. The weights of the individual criteria are then determined by:

- Determine the standard weights of the individual groups of criteria (by using some of the above-mentioned methods),
- Determine the standard weights of each criterion in the relevant group,
- By multiplying the weights of the criteria groups and the weights of each criterion within each group we will find the resulting standardized weights of the criteria.

3.2 Methods of Determining the Order of Variants

The aim of the multi-criterial evaluation of the variants is to determine the ranking of the advantages of each variant in terms of the chosen criteria, with the best placement option being the best compromise variant. The methods for choosing a compromise variant between non-dominated variants differ from the approach to the "compromise variant", the difficulty, and the usability of different types of multi-criteria tasks. The results obtained by various methods are therefore subjective and may differ from one another.

Methods can be divided according to what type of information they require.

– Methods requiring knowledge of aspiration level of criteria values—this group of methods include, for example, the *conjunctive method, the disjunctive method* and the *PRIAM method*. Information on the importance of the criteria is expressed as the aspiration level of the criteria. Criteria values of all variants with aspiration levels of all criteria are compared. Typically, the group of variants is divided into two groups:

  – variants that have worse critical values than the aspiration level set (unacceptable, ineffective),
  – variants that have better or the same critical values than the aspiration level (acceptable, effective). With sufficient tightening of the aspiration levels, the only option left in the set of acceptable variants can be called compromising.

– Methods requiring ordinal information on variants according to each criterion—for example, *ordering method, lexicographic method, permutation method, ORESTE method*.
– Methods requiring cardinal information on variants according to each criterion—this group of methods is further subdivided into subgroups according to the principle on which the evaluation is based. There are these basic approaches:

  – maximization of utility (*weighted sum method, basic variant method, Analytical Hierarchy Process method, scoring method*),
  – minimizing the distance from the ideal variant, or maximizing distance from basal variation (*TOPSIS—Technique for Order of Preference by Similarity to Ideal Solution*),
  – preferential sessions (*ELECTRE—ELimination and Choice Expressing REality, PROMETHEE—Preference Ranking Organization Method for Enrichment Evaluation*),
  – methods based on the limit of substitution (*the method of successive substitution*).

Some of the mentioned methods will now be discussed in a greater detail.

Conjunctive and Disjunctive Method

Conjunctive and disjunctive methods are especially recommended for "pre-selection" of variants, which are then evaluated by other methods. This means that aspiration levels are not set so strictly that only one solution is suitable. A group of variants that suits aspiration levels is sought for. One could say that these methods will remove the badly and unsuitable variants and the others will then be evaluated by other methods.

When applying these methods, it is necessary that the aspiration levels of all the criteria and the cardinal evaluation of the variants according to the criteria are known. According to the aspiration level we divide the variants into acceptable and unacceptable ones. In the case of a conjunctive method, we only allow variants that meet all aspiration levels.

Assuming that all criteria are maximizing, only variants A_i which, according to all considered aspects, have a value of a predetermined aspiration level z_j or greater value will be in the set S of acceptable variants. Therefore:

$$S = \left\{ A_i : y_{ij} \geq z_j, j = 1, 2, \ldots, n \right\}.$$

Assuming that all criteria are minimizing, then in the set of acceptable variants S there will be variants, which, according to all considered aspects, have the value of the predetermined aspiration level z_j or lower value. Therefore

$$S = \left\{ A_i : y_{ij} \leq z_j, j = 1, 2, \ldots, n \right\}.$$

In the case of a *disjunctive method*, we admit variants that meet at least one requirement. In the case of maximization criteria, it true that from all possible alternatives A_i, in the plurality of acceptable variants S, there will be those variants which at least according to one considered aspect have at least a predetermined aspiration level value or greater. Mathematically we can write:

$$S = \left\{ A_i : \exists\, j = 1, 2, \ldots, n, y_{ij} \geq z_j \right\}.$$

In case of minimization criteria

$$S = \left\{ A_i : \exists\, j = 1, 2, \ldots, n, y_{ij} \leq z_j \right\}.$$

If the requirements expressed by aspiration levels are too strict, the set of acceptable variants is empty. In such case, it is necessary to specify new, moderate aspiration levels. Conversely, if the requirements are mild, the set of variants will be too large. It is then necessary to tighten the aspiration levels.

Analytical Hierarchy Process

When solving decision-making problems, it is necessary to take into account all the elements that influence the outcome of the analysis, the links between them and the intensity with which they interact. The decision problem can be recognized as a hierarchical structure. It is a linear structure containing several levels, each of which includes several elements. The layout of each level is always from general to specific. For the general role of multi-criterial evaluation of the variants, the following hierarchy can be used: 1st level—negotiation objective; 2nd level—experts involved in the evaluation; 3rd level—evaluation criteria; 4th level—the variants under consideration.

Similarly, as among the criteria for weighting the Saaty method's criteria, relationships between all components at each level of the hierarchy can be determined. If we have a four-level hierarchy, i.e. one objective, h experts, n criteria and m variants, the second level of the hierarchy will be a pair of matched matrices of $h \times h$. On the third level, the matrix $n \times n$ and the fourth level of the matrix of dimensions $m \times m$. Using calculations (see Saaty's method for calculating criteria weights) in these matrices, the variants "divide" the weight value of the relevant criterion (the criteria then "divide" the weight of the relevant expert). The values we get are called pre-fringe variance indexes in terms of all criteria. Therefore, if we add these preferential indices in terms of all criteria, we will obtain a variant assessment from the point of view of all the experts and from the point of view of all criteria. A detailed description of the other methods can be found in literature, e.g. Saaty & Vargas (2000), Skulinova (2013) and Lacko (2008).

3.3 Sensitivity Analysis of Preferential Order of Variants

The ranking of the advantages of the given decision-making variants determined by one of the multi-criteria assessment methods of the variants depends mainly on the weights of the individual criteria and on the method used.

Examination of the sensitivity of the preferential order of the variants to the determination of the importance of the individual criteria falls within the field of experimentation on models where the multi-criteria evaluation of the variants is carried out under the changing weightings of the criteria. Only when, in these changes, the preferential order of the variation (or at least the highest valuation variant) does not change, this order is not sensitive to the inaccuracy of the weights and it can be said that the importance of the individual criteria has been correctly assessed by the decision maker. If the preferential order of the variants is highly sensitive to the weighting of the criteria, their reliability must be increased.

The dependence of the preferential order of the variants on the method used for their multi-criteria evaluation is determined by the fact that the different methods are based on different, usually simplifying assumptions, and that, due to the different approach to the term "compromise variant", they use different computational processes. Therefore, it is recommended to use multiple methods in multi-criterial evaluation of variants and to verify the sensitivity of the preferential order of the variants with respect to the methods used. Only the option that remains in the first place using any method can be considered the most advantageous.

3.4   Decision-Making with Uncertainty Using Multi-criteria Analysis of Variants

In order to make the best decision when uncertain, usually a decision-making model based on a model of a game with nature is used. The applicant chooses strategies—alternatives whose resulting effect is influenced by the state of circumstances.

The state of uncertainty means that the applicant does not know how the situation will develop in the future, does not have any information which of the possible states of circumstances will become true. For solution of such a model various procedures developed in the theory of games is used, e.g. maxima or minimax principle, Laplace principle of insufficient reason, Savage's or Hurwicz's principle. All of these cases when selecting the best alternative of a decision lead to negligence or even rejection of some results of their realization.

References

Lacko, B. (2008). Methods and techniques of project management. In *Proceedings of selected chapters from project preparation and management. Educational (In Czech)*. Novy Jicin: Regional educational facilities for pedagogical education employees and information center. Retrieved 2017, from http://www.logframe.cz/metoda.htm.

Saaty, T. L. (1980). *The analytic hierarchy process*. New York: McGraw-Hill.

Saaty, T. L., & Vargas, L. G. (2000). *Models, methods, concepts and applications of the analytic hierarchy process*. Boston: Kluwer Academic Publisher.

Skulinova, D. (2013). Multi-criterial decision making in determining the usability of building in areas with increased industrial activity (In Czech)). Brno. Retrieved 03 27, 2013, from https://is.mendelu.cz/eknihovna/opory.

Chapter 4
The Use of Multi-criteria Analysis in Evaluation of Utility Value DGI

Multi-criteria analysis can be used to evaluate functionality of digital data or to analyze geographic conditions in a given space. The following text describes the use of multi-criteria analysis for improving the utility qualities of DGI.

4.1 Understanding of Utility Value

The utility value of a product, service, etc. may be detected with an advantage using the method of value analysis which was developed in detail by Miles (1989). Value analysis always primarily tries to reach maximal functionality, i.e. optimal satisfaction of social needs by the given product.

Product functionality may be expressed as function F or a set of functions F_i which express concrete requirements of users, i.e. they express utility value of a product or functionality. To secure the mentioned functions it is necessary to spend expenses E or E_i. Expanses do not have to be expressed only in monetary units, but they can be expressed in units of time, amount of work, number of people involved, etc.

In decision-making processes various options of the required products are usually considered. They differ from each other in functionality as well as in expenses necessary for provision of such functionality. The ratio between functionality and expenses can be optimized by methods of utility analysis and in order to find the best of j options in that particular case. The aim of each case of use of utility analysis is to maximize the ratio between functionality and expenses to its provision. Functionality F_j of the j variant may be expressed in Eq. (4.1):

© Springer International Publishing AG, part of Springer Nature 2019   49
V. Talhofer et al., *Quality of Spatial Data in Command and Control System*, Studies in Systems, Decision and Control 168, https://doi.org/10.1007/978-3-319-94562-0_4

$$F_j = \sum_{i=1}^{n} F_{ij,} \qquad (4.1)$$

and similarly, also expense of the j variant Eq. (4.2):

$$E_j = \sum_{i=1}^{n} E_{ij}. \qquad (4.2)$$

The indicator of *relative cost effectiveness* RCE_j is the criterion of efficiency, it is given by the *relation in the indication* of (4.3):

$$RCE_j = \frac{\sum_{i=1}^{n} F_{ij}}{\sum_{i=1}^{n} E_{ij}}, \qquad (4.3)$$

where

– i is 1., 2., 3., … n function,
– j is 1., 2., … m variant.

Functionality of a product is evaluated in relation to users. Methods of value analysis examine its restricted behavior and side effects. The basic procedure is *function analysis* which concerns development tendencies in behavior of purpose system and changes in its structure that occur in the process of realization of its utility value. Apart from this, function analysis is used to find out about expenses needed for the provision of the required level of all functions of the analyzed product.

Function analysis is based on a thorough analysis of the core of the product and manifestations of its behavior. The process of transformation of a concrete form of the analyzed product into its general replaceable form with the help of a set of functions expressing concrete users' requirements is the methodical core of the function analysis. As most of such defined functions may be fulfilled in various ways which manifest in various variants of the product, it is possible to optimize the resulting function and together with optimization of expenses to maximize the proportional effective value *RCE*.

Functions expressing users' requirements are expressed by characteristics which is a detection and quantified expression of technical, utility, emotive, expressional, and economical features (parameters, values, characteristics) that are the base for their evaluation. Function analysis detects values expressing:

– Function specification,
– Function significance,
– Degree of fulfilling of functions,
– Expenses for provision of the function.

4.1.1 Specification of Functions and Their Significance

Specification of a function is understood as a provision of functionally conditioned characteristics which as technical, utility or emotive features complement the description of functions. They are actually *criteria* whose measurable or other determinable values must be maintained so that the product functions and fulfils its functions on the required level. With the help of parameters, values and characteristics of these functionally conditional characteristics, the degree of completeness of product function is indirectly expressed.

Function significance is a value that expresses the importance of the function with regards to the whole set of functions. The value of function significance is expressed in units of significance. It is calculated from the complete value of significance of the set of functions of the product which is constructed as cardinal scale of a hundred- or thousand-scale of units of significance. The value of significance of one function from the complete value of significance of the complete set of functions is determined as a share whose size depends on the value of classification coefficient of function significance—or coefficient of significance (p_i)—and the range of the scale of significance units. The coefficient of significance is a number whose growing size expresses a higher importance, weight of the analyzed function, or functionally conditional feature. Weight of functions or functionally conditional features may be set by a logical consideration, in more complicated cases then by the method of pair comparison or direct estimation in the given classification scale.

Firstly, for determination of the value of significance of a function, a logical consideration, for more complicated cases, *the method of pair comparison* determines the order of importance of all functions. Then, using *the method of proportional numbers* or *gradual weighting*, the coefficient of function significance is calculated and finally with the help of *the method of deviation scale* the values of significance of the individual functions are determined.

The method of pair comparison lies in a consequent comparison of importance of one function with the importance of all other ones and determination of which of them is the most important. This method in general can be used if it is possible to expect a regular interval of distribution among coefficients of significance of the individual functions.

The method of proportional numbers follows the determination of the ratio of importance of the evaluated function with regards to the least important function whose coefficient of significance is set to 1. This method is suitable in cases when there are expected big differences in the importance of the individual function.

The method of gradual weighting is used in cases when functions are divided into logical groups. In such case, firstly scales (such as percentage share) of the individual logical groups are set, e.g. using the method of pair comparison. These weights are then budgeted to the individual functions of the appropriate logical groups again in percent within one logical group. Out of them, the final percent weights set within the complete set of functions are calculated. In general, the method can be used also for multi-level division of functions.

Apart from the mentioned methods of determination of value of function significance, it is possible to use a questionnaire survey as well and then use statistical methods for its evaluation. However, when using questionnaire survey, it is also possible to use the above-mentioned methods, such as pair comparison.

4.1.2 Degree of Fulfillment of Functions

The degree of fulfillment of a function is a number data which expresses the level of fulfillment, exceeding or not reaching the needed optimum of functionality of the given function that is realized by the evaluated product. One of the basic information for setting the degree of fulfillment of a function is the values of functionally conditional characteristics, i.e. function specification. If the function has only one characteristics, it is possible to evaluate the degree of its fulfillment by one quantitatively measured value. In most cases—which also evaluation of utility value of digital geo-information belongs to—however, it is functionality created by several measurable as well as non-measurable values and characteristics that express the degree of *fulfillment* of functions and *functionally conditioned characteristics*. In such cases for expression of the given degree of fulfillment of functions, evaluation scales are used—nominal, ordinal, and cardinal scales.

Nominal scale is two-value that evaluate functionally conditioned characteristics in the form of "yes/no", "comply/not comply", "0/1", etc.

Ordinal scale by its number only defines the order of functionality of the evaluated variants. It does not set by how much the functionality of one variant is higher or lower from another variant. As a variant of the ordinal scale, it is possible to apply *point* word-numerical *scale* where a certain word characteristic of degree of fulfillment of functionality corresponds to a certain number of points.

Cardinal scale defines in numbers the mutual ration of functionality of the individual evaluated variants, i.e. how many times, or by how many numerical units the functionality of one variant is bigger or smaller than the other. The level of functionality on cardinal scale is derived from a verbal evaluation of qualitative characteristics as well as from exactly measured values of quantitative criteria. The most frequent form of a cardinal scale is a numerical scale of decimal system.

Numerical scale of decimal system is divided into a coherent number of points—degrees. The most frequent are ten-, hundred-, or thousand-point scales. For practical usage it is recommended to use a hundred-point scale which is analogic to percentage points, when the optimum is expressed as 100% fulfillment of the required functionality. Percentage number is also used for a transfer of really detected values of quantified parameters and verbally expressed statistics of qualitative criteria to relative values of a cardinal scale. This procedure enables to transfer all data to *a unified evaluation scale*.

It is possible to use several methods in order to set a degree of fulfillment of functions. Their choice depends on the character and number of criteria and their parameters and to the significance of the individual functions.

The *exact method* can be used in situations where it is possible to exactly measure the value of the given parameter of a function or criteria for its fulfillment. In case of geo-information, these are e.g. coordinates of position and height, area, etc., if these values also express functionality. However, if it is necessary to use at least one more parameter or qualitative data to express functionality, the exact method changes to point or classification methods.

When the significance of evaluated functions equals, it is possible to use *point method*, with which the degree of function fulfillment is expressed by point value, most frequently ordinal scale. In order to increase the objectivity of results, it is possible to use also one of cardinal scales. The degree of fulfillment of a given function is then defined by such point value which expresses how each of the variants in relation to the defined optimum (the highest point value, the highest value of cardinal scale, etc.) fulfilled the given function. The highest sum of points gained from the evaluation of all functions then represent the highest level of fulfillment of the complete functionality of a product.

With various significance of functions, it is necessary to choose a *classification method* in order to define the complete functionality of j-variant of a product (object). Its application needs four steps:

1st step Using the method of pair comparison, the order of importance of functions is set, and using a suitable method, coefficient of its significance (weight) p_i is set.

2nd step The point method evaluates and in criteria k_{ij} expresses how the analyzed object fulfills the given (i-) function in relation to social optimum.

3rd step The resulting value of degree of fulfillment of one function is set as a product of coefficient of its significance and the point value of fulfillment of criteria according to relation:

$$F_{ij} = p_i k_{ij}. \tag{4.4}$$

4th step The resulting degree of fulfillment of all n functions of the set which as a whole characterize the complete functionality of the analyzed object are set by a suitable aggregate function. Aggregation function is usually chosen as a simple additive function solved according to relation:

$$F_j = \sum_{i=1}^{n} F_{ij} = \sum_{i=1}^{n} p_i k_{ij}. \tag{4.5}$$

However, the disadvantage of this function is that it assumes mutual independence of the individual functions, which does not always have to be true with geo-information products. For example, it is not possible to express such a state (result of evaluation) with this function when the functionality is objectively completely devalued due to non-fulfillment of only some of the criteria. Such situation may be better expressed by a product aggregation function in the form:

$$F_j = \prod_{i=1}^{n} F_{ij} = \prod_{i=1}^{n} p_i k_{ij}. \qquad (4.6)$$

Characteristics of the stated function is that each change of evaluation of a partial function is in the same ratio transferred also to the overall evaluation. In this sense it is absolutely direct dependence of the individual functions among each other. However, not even this condition does not have to be completely met and there are cases when a part of functions is mutually independent and a part is dependent. Then the aggregation function F may be written in a general expression:

$$F_j = \prod_{k=1}^{r} F_{kj} \sum_{l=1}^{s} F_{lj} = \prod_{k=1}^{r} p_k k_{kj} \sum_{l=1}^{s} p_l k_{lj}, \qquad (4.7)$$

where

– n is the number of all functions in the set,
– r is the number of dependent functions,
– s is the number of independent functions ($s = n - r$).

If the individual parts of one product are evaluated and they are created according to the same project, but each part has other characteristics expressed by the level of fulfillment of the individual evaluation criteria, it is possible to express also the functionality of the individual partial parts with the help of the *individual utility value*. When evaluating utility of these individual parts, it is good first to define the *ideal level of quality* which then serves as a comparison standard for expressing the level of fulfillment of the individual criteria u_s. With the help of comparison standard, it is then possible to evaluate the level of fulfillment of the individual criteria and consequently also the complete utility value.

The level of fulfillment of the individual criteria in general may be expressed as:

$$u_s = \frac{k_s}{k_s^*}, \qquad (4.8)$$

where

– k_s is the value of fulfillment of s-partial criterion,
– k_s^* is the level of fulfillment of s-partial criterion or criterion of its sub-group of the comparison standard.

The level of fulfillment of the individual criteria shall be calculated with regards to their structure, i.e. to their division into partial criteria.

The complete individual utility value of a part of a product (its functionality) is then given by the aggregation function of the same type as one of the above-mentioned aggregation functions suitable for the given type of the product. So for a general expression of an individual utility value it is possible to write:

$$F_j = \prod_{k=1}^{r} F_{kj} \sum_{l=1}^{s} F_{lj} = \prod_{k=1}^{r} p_k u_{kj} \sum_{l=1}^{s} p_l u_{lj}. \tag{4.9}$$

4.1.3 Expenses for Functions

Expenses for functions are usually a quantity that has an objective base *in the form of monetary units* in which necessary expenses for work to secure the analyzed function are expressed. They are usually secured by detailed analyses of salary and material expenses in a production organization which provides the evaluated product.

Expenses for functions, however, do not have to be expresses solely in monetary units. Especially in case when it is necessary to secure maximal functionality of a product in a limited time and restricted conditions of personnel security of creation or updating a product, expenses for functions may be expressed by *time expenses*. This is a frequent case when solving critical situations when security of the needed product is not a question of monetary units at disposal, but above all the questions of time during which it is necessary to create or update the given product. Then it is about setting priorities and a subsequent order of solutions of the individual steps in a way so that also in restricted time conditions maximizing of functionality is secured.

When determining the exact expenses of the individual functions of a product, in practice of value analyses, *the analytical-calculation method* was applied. In cases when it is not necessary to determine expenses of the individual function in detail, it is possible to use *the method of expert estimation*. This method lies in several steps:

1st step A list of all elements (parts, operations, etc.) which create the structure of a product and giving their values expressed in expense units. The listing may be done in a vertical column of an evaluation table.

2nd step All functions of the analyzed product are written horizontally.

3rd step A check of completeness of the composition of functions is performed by evaluating whether each element of a product may be assigned some of the functions or functionally conditioned characteristics. If it is not the case, then it is examined what function it actually has, whether it is a function of the analyzed system and not only of its element. Missing functions of the product as a system are added. At the same time, it is evaluated whether unnecessary functions may be found in the evaluation table.

4th step Elements of the product are finally assigned to its functions.

5th step By expert evaluation, shares of elements of a product on fulfillment of the individual functions are assigned.

6th step Percentage differences are used to calculate expenses of elements on fulfillment of the individual functions.

7th step The sum of values of expenses in the individual columns gives the expenses of the individual functions. The total sum of expenses of functions and the total sum of expenses of elements must be equal.

When examining the established products of geo-informatics, however, it is necessary to modify this procedure. Spent expenses may be examined from technological sheets in which the individual stages of preparation, creation and updating of products with high accuracy are normalized also with respect to the difficulty class of the particular storage unit given by geographic and non-geographic conditions of the terrain modelled in this storage unit.

4.1.4 Use of Value Analysis

A simple demonstration of the use of a value analysis for solution of a choice of a suitable variant of a product is documented in the following general example. Let us have 4 variants of one product whose functionality and expenses are stated in Table 4.1. Limit expenses given by the recipient are given by the value of 14.55 expense units and limit functionality by the value 17.20.

If the stated values are shown in a graph (Fig. 4.1), it is obvious that none of the variants of the product complies. Variants 2 and 3 do not reach the required functionality and variants 1 and 4 are above the limit of expenses. So now it is up to the recipient to decide whether they are willing to pay higher expenses for variant 4 which meets their requirements and whose price exceeds the limit only by 2.45 expense units, or whether they will ask the provider to modify this variant to decrease the expenses, yet not functionality.

The given example is only illustrative, without considering the way of determination of the utility value—functionality, neither expenses for creation of such product. Possibilities of application of this theory for evaluation of quality when working with digital geographic information aimed at decision-making processes is discussed in the following paragraphs.

Table 4.1 Functionality, Expenses and Relative Cost Effectiveness of a product (*Note* the stated values are only illustrative)

Variant	Functionality	Expenses	Relative cost effectiveness
Variant 1	29	27	1.07
Variant 2	11	9	1.22
Variant 3	14	18	0.78
Variant 4	25	17	1.47

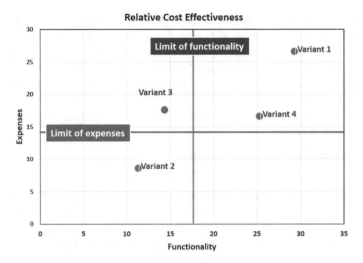

Fig. 4.1 Graph of distribution of expenses and functionality of variants of a product

4.2 User Functions of DGI and Methods of Their Evaluation

Chapter 1 specified 6 basic functions of digital geographic information with their detailed definition. There were the following functions defined:

- Information function,
- Model function,
- Function of source for mathematical modelling, projecting and planning,
- Function of means of automatization,
- Illustration function,
- Function of source for derivation of other DGI.

  According to methodology of evaluation analysis, for each specified function it is necessary to determine its significance for solution of the given problem or a group of problems. Furthermore, it is necessary to specify functionally conditioned characteristics—criteria that may be quantified or classified and that can be used for evaluation of functionality of the individual products, their parts or their variants.

4.2.1 Evaluation of Function Significance

Defined functions have different significance for each user. This significance can be evaluated by weight expressing the degree of need of the stated function of the given product. It is good to do quantification of function significance by user research with chosen users who know the issue of DGI and their usage very well. Basically, it does not make sense to do this research among end users who use the products only on and off and usually do not think about the actual characteristics of data. Procedures from the environment of the Army of the Czech Republic (ACR) and its Geographic service were used for documentation of the methodical procedure of the complete value analysis. The stated procedures are fully applicable also for the field of general crisis management.

Quantification of significance of functions was performed by research within users and producers of DGI in two independent ways:

1. Pair comparison of significance of functions,
2. Direct estimation in the given classification scale.

For the research, chosen standard products DMU25, DMU100, MGCP, DMR4, SRTM were used. They are used within ACR, Fire Rescue Service, or police units.

Digital Model of Territory 25 (DMU25) is defined as a vector representation of physical-geographic objects from the territory of the Czech Republic and close territory abroad with information detail corresponding at least to topographic maps of 1:25,000. Geometric representation of each type of object that is mentioned in the Catalogue of Topographic Objects (CTO 25) (MoD-GeoS, 2013) has its own defined class of elements in the geodatabase. The whole geodatabase contains in total 180 classes of elements in thematic layers:

– Hydrology,
– Transport network,
– Pipelines and energy,
– Vegetation cover,
– Built-up area (residence),
– Borders,
– Altimetry (relief).

In meta-data information each object contains quality of delimitation of horizontal accuracy ACC (Accuracy Code) where information whether an object complies with the criterion of product accuracy, exceeds given tolerance, horizontal delimitation is unsure or it is more accurate than product requirement. Furthermore, also code RDV is stated here, it informs about the date of last revision. The catalogue of objects conforms to international standards MIL-STD-600006: Vector Product Format (DoD, 1992) and DIGEST-1-2: Digital Geographic Exchange Standard (DGIWG-103, 2008).

Digital Model of Territory 100 (DMU100) is a similar product to DMU25. Its vector database contains the same thematic layers and basically the same objects (MoD-GeoS, 2006). However, the choice of objects and their horizontal delimitation is generalized and it corresponds to topographic maps of 1:100,000. This model conforms to the same standards as DMU25.

Product *MGCP* is a result of international project called Multinational Geospatial Coproduction Program that 30 countries work on. Database DGI is created in basic units $1° \times 1°$. Geometric and thematic level of detail corresponds to topographic maps of 1:50,000 or 1:100,000, their quality follows the possibility of the producer of the given storage unit. The declared mean positional error is up to 25 m. A detailed description of the product is given in (MGCP, 2012). Basically, the database contains the same objects as DMU100 and conforms to the same standards as well.

Digital model of relief of the Czech Republic of 4th generation (DMR4) models earth surface in the form of heights of discreet points in a regular net of 5×5 m that have coordinates E, N, H.

Coordinates E, N are planar coordinates in geodetic reference system WGS84 and projection UTM, H represents altitude in the height reference system Baltic Vertical Datum—After Adjustment with a total mean error of height of 0.3 m in open terrain and 1 m in wood terrain. The model was created from data gained by the method of laser scanning of altimetry of the Czech Republic.

SRTM (Shuttle Radar Topographic Mission) is a digital model of relief generated from data gained during the mission of space Shuttle in 2000 and after specification in 2015 (JPL, 2015). Again, it is a regular net of height points of almost complete surface of the Earth, excluding polar areas, in a fixed interval set in geographic coordinates in geodetic reference system WGS84. In the USA, the grid is 1×1 arc-second, other areas 3×3 arc-second. Heights are given in height reference system EGM96. The area of the Czech Republic gets an interval in position of approximate size of 60×90 m. The accuracy in setting the height on the area of the Czech Republic is about 16 m and it is independent on covering the relief with other objects.

The research itself was conducted with the help of questionnaires with 18 respondents who mostly belonged to Geographic Service of ACR or they were already experienced with work with DGI in their own expertise, e.g. in tactics and combat use of artillery. The respondents from the Geographic Service were chosen so that they had extensive experience with using digital geo-information in army practice. That is the reason why they were chosen first of all from units of geographic support of operational headquarters.

The respondents set weights of functions in pair comparison as well as in direct estimation in classification scale 1,...,10, where value 1 was chosen for the lowest significance and value 10 for the highest significance. The individual weights were in both cases recalculated so that their sum equals 1. The resulting weights of functions (Table 4.2) were calculated as an average value from values of both methods.

Table 4.2   Weights of functions gained by pair comparison and direct estimation in a classification scale

	DMU25	DMU100	MGCP	DMR4	SRTM
Information	0.159	0.204	0.202	0.115	0.167
Model of reality	0.222	0.186	0.149	0.221	0.162
Source for mathematical modelling	0.195	0.189	0.152	0.206	0.159
Mean for automation	0.141	0.162	0.166	0.162	0.140
Illustration	0.069	0.132	0.212	0.139	0.229
Source for derivation	0.214	0.127	0.119	0.157	0.143
Sum	1.000	1.000	1.000	1.000	1.000

4.2.2   Functionally Conditioned Characteristics of a Digital Model of Territory, Criteria of Their Evaluation

Digital geo-information reflects projected and described reality in a model that is defined based on user requirements and created with the help of tools for modelling, especially with the help of abstraction and generalization. A model created this way may never be an exact copy of reality, yet in an ideal case it should fully satisfy most users. However, with regards to variedness of users and their needs and technical, technological and capacity possibilities of model producers, the mentioned ideal state basically may never become true. A responsible model producer will always strive to secure maximum user functions, i.e. maximal functionality of the model or its maximal effectiveness. For expression of the complete functionality, however, it is not possible to use the individual functions of DGI as they are mentioned in the previous paragraph, because these functions are impossible to be measured scientifically. That is why it is necessary to use functionally independent and measurable, or any other way quantifiable criteria for their expression.

4.2.2.1   Qualitative Characteristics of Digital Geo-Information

When setting criteria for measurement of the utility value, it is good first to follow the list of features and characteristics that are most often mentioned as significant or required, from the point of view of the producer as well as their user. The following characteristics are either exactly measurable, comparable or evaluable, or they are of such character that their evaluation may be done only by experts based on expert estimation or questionnaire survey among users. Based on research in literature and an additional survey among users, the following list of qualitative and quantitative characteristics of DGI was created:

– Completeness of content according to norms for data processing,
– Completeness of content according to user's needs,
– Range of major, minor, and formal defects,
– Completeness of addition of qualitative and quantitative characteristics of saved objects and phenomena,
– Correctness and unambiguity of qualitative and quantitative characteristics of saved objects and phenomena,
– Unambiguity of localization of qualitative and quantitative characteristics of objects and their geometric shapes,
– Compliance with the prescribed spatial data accuracy,
– Compliance with the prescribed attribute data accuracy,
– Compliance with the prescribed time data accuracy,
– Compliance with the prescribed spatial resolution of data,
– Compliance with the prescribed time resolution of data,
– Compliance with the prescribed attribute resolution of data,
– Compliance with the prescribed classification of objects and phenomena,

– Creation of correct topology of saved objects,
– Compliance with alignment of objects and phenomena on the edges of work spaces,
– Up-to-dateness of data in regards to the state in terrain,
– Absolute number of changes that occurred in terrain in the period of time since the last data version,
– Significance of an area modelled in the database for a user,
– Economic and political significance of the area modelled in the database,
– Completeness of the database of the area of interest,
– Complete technical level of the database,
– Data consistency in the database and their maintenance in process of updating,
– Usability of data with regards to a user's needs,
– Independence of data on program environment,
– Data standardization according to military standards,
– Data standardization according to civilian standards,
– Compliance of the prescribed principles of standardization,
– Security of data against abuse,
– Security of data against unintentional damage or destruction,
– Suitability of data to be used for modelling and simulations, or the possibility to separate from them or create suitable data for modelling and simulations,
– Suitability of data to be used in controlling and dispatching systems,
– Knowledge and transparency of the used source materials,
– Knowledge and transparency of the used methods and mathematical models when using the source materials,
– Knowledge and transparency of the used methods and mathematical models when deriving secondary data from the primary sources.

The given list is certainly not complete but it reflects the main basic characteristics of DGI. The list may be reduced or on the contrary extended according to types of tasks and activities in which they work with DGI.

4.2.2.2   Definition of Functionally Conditioned Characteristics of Digital Geo-Information

From the above-mentioned qualitative characteristics of DGI we can make the following five generalized basic functionally conditioned characteristics—criteria:

1. *Content of database*, which evaluates above all the conformity of its definition and user's needs. It expresses the relationship between the real, modelled world and objects and phenomena saved in the database. It does not evaluate the completeness of the individual saved objects and phenomena and their characteristics, but whether the given norms and regulation for creation of databases which reflect characteristics of the created terrain model involve all objects and phenomena needed by the user, if these elements are correctly defined geometrically as well as thematically.

2. *Technical quality of database content* which evaluates above all the quality of saved data. The criterion is complex and it involves an evaluation of source materials and methods of their transformation, positional and attribute accuracy of data, its logical consistency and complexity of the saved objects and phenomena.

3. *Database timeliness* expresses the condition of the database in regards to current date and potentially also the speed of its outdating. In principle it enables to evaluate currentness of the individual groups of content elements as well as homogeneity of currentness of the complete database and system of its maintenance.

4. *Area importance* is given by users' needs so that it fulfills their requirements on a spatial range of the processed or secured territory. Significance of this criterion is given with regards to the fact, that for each user the territory has a different meaning, e.g. military, political, economic, etc. Databases containing data from various territories are not in any case mutually substitutable. It is necessary to mention here that if there is a limited spatial range of territory, each part of the territory is not of the same significance for every user. E.g. for defense of the country against external threats, border areas are of greater significance than inland areas; surroundings of large inhabited agglomerations are of higher significance in terms of extraordinary traffic events than scarcely inhabited areas with minimum of communications, etc. It is also necessary to consider that absence of data from certain areas may lead to restricted functionality of the whole database.

5. *User friendliness*. This criterion is understood as the ability of data to be used in various types of program environments, be it of GIS character or various application programs that are reflected first of all in the question of keeping standardization principles. Also, independence and data security are evaluated in this criterion. Independence of data expresses above all program independence so that it would be possible to use them without any modifications in various program environments and secure their full usability. Data safety represents a system of data security against unintentional or intentional damage, abuse, or loss.

Based on evaluation of fulfillment of the mentioned criteria, it is possible rather reliably to define a utility value of the individual products. Weight is assigned to each criterion corresponding to its significance for the calculation of the complete functionality of the product.

The Table 4.3 shows weights of the individual criteria for the above-mentioned products. These weights were surveyed in the same way which was used when setting weights of the individual functions. In the first phase, the weights were set by a pair comparison and in the second phase they were again directly estimated in classification scale 1,…,10. Table 4.3 contains an average value of weight of the criteria from both methods together with their reclassification so that their sum for one product equals 1.

Weights of criteria, however, must be adjusted based on their significance for the individual functions. Table 4.4 is the result of survey where respondents set how significant the individual criteria are for the provision of fulfillment of the individual

Table 4.3 Average value of weight of criteria

Criterion	DMU25	DMU100	MGCP	DMR4	SRTM
Content of database	0.240	0.230	0.215	0.209	0.197
Technical quality of database content	0.200	0.189	0.175	0.216	0.189
Database timeliness	0.220	0.197	0.179	0.180	0.215
Area importance	0.199	0.230	0.230	0.233	0.229
User friendliness	0.141	0.154	0.201	0.162	0.170
Sum	1.000	1.000	1.000	1.000	1.000

functions. The average weight of criterion was calculated from the sum of point evaluations of the individual respondents.

At the end, the resulting values of weights of the individual criteria were calculated as average values from data from Tables 4.3 and 4.4. The resulting values are shown in Table 4.5.

The resulting weights may be documented also graphically (Fig. 4.2).

It is obvious from the graph that the highest significance carries criterion "content of database" basically in all evaluated products. Criterion "user friendliness" gets quite small weight the reason for which may be that nowadays a lot of systems for work with DGI are able to accept also data in various standards.

Resulting weights p_{ij} from Table 4.5 are for each product recalculated for partial criterion. The example for DMU25 is given in Table 4.12.

The level of fulfillment of the individual criteria is evaluated based on measurable values or expert estimations. The next paragraphs speak about the methodology of this evaluation.

Contents of Database

As said before, the criterion *contents of database* expresses first of all conformity of the defined contents in regards to user's needs. These user's requirements must be accepted first of all when outlining the given model, i.e. when creating a conceptual and logical model of the considered database. All required objects and phenomena of the modelled reality must be defined in these models and at the same time, they must have a corresponding positional and thematic level of detail given especially by the used procedures and tools of abstraction and generalization which are used to model "The real world".

The base for the conceptual and logical model of the given database is the norms and regulation for its definition as a physical model and then technological norms or instruction for the entering of data itself. A catalogue of objects that contains definitions of all objects and their characteristics is always a part of the mentioned documentation. Such catalogues are e.g. catalogue of objects of DMU25 (MoD-GeoS, 2013) or MGCP (MGCP, 2012).

Table 4.4 Significance of criteria for function

Criterion	Function							
	Information	Model of reality	Source for mathematical modelling	Mean for automation	Illustration	Source for derivation	Sum of evaluation	Mean weight of criterion
Content of database	130	114	118	98	97	108	665	0.226
Technical quality of database content	86	112	101	114	82	101	596	0.203
Database timeliness	93	120	120	130	80	101	644	0.219
Area importance	111	92	102	91	94	89	579	0.197
User friendliness	80	72	73	81	76	74	456	0.155
Sum	500	510	514	514	429	473	2940	1.000

Table 4.5 Resulting values of weights of criteria

Criterion	DMU25	DMU100	MGCP	DMR4	SRTM
Content of database	0.233	0.228	0.221	0.217	0.211
Technical quality of database content	0.201	0.196	0.189	0.209	0.196
Database timeliness	0.219	0.208	0.199	0.200	0.217
Area importance	0.198	0.213	0.214	0.215	0.213
User friendliness	0.148	0.155	0.178	0.159	0.162
Sum	1.000	1.000	1.000	1.000	1.000

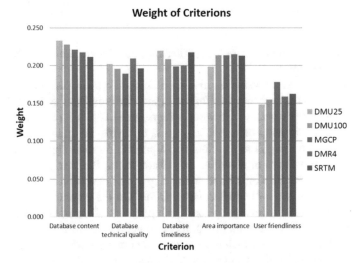

Fig. 4.2 Graphic expression of the resulting values of weights of criteria

That is the reason why criterion contents of database evaluates conformity of norms and regulations for the creation of database and catalogue of objects with user's requirements and also their respecting when processing of the database. The whole product is evaluated by this criterion.

The criterion is further divided into two partial criteria of 1st group and two partial criteria of 2nd group. Its structure is given in Table 4.6.

Criterion k_{11} *completeness of real world model* belongs to the first group. Here it evaluates the conformity of the created model with a summary of user's requirements on the model that are given by their needs and needs of user's applications. The value of the criterion is expressed by relation:

$$k_{11} = 100 - \alpha_{11}, \tag{4.10}$$

Table 4.6 Structure of criterion *content of database* (*Note* this table is a part of the Table 2.2)

Group of main characteristics—main criteria	Characteristics of partial criteria	Definition	Quality criterion
Content of database—k_1	Completeness of real world model—k_{11}	Conformity of the defined model with users' requirements	Percentage rate of incomplete information
	Compliance of required resolution for geometric and attribute data—k_{12}	Conformity with the required level of detail of data	Percentage rate of objects with not fulfilled conditions of geometric level of detail —k_{121}
			Percentage rate of objects with not fulfilled conditions of thematic level of detail —k_{122}

where α_{11} is a value in the range 1, …, 100 expressing the degree of nonconformity with user's requirements.

It is possible to set the value α_{11} based on user's survey in which the individual users comment on the degree of fulfillment, or non-fulfillment, of their requirements on the set contents of database. However, with regards to the fact that GDI is usually created for a larger range of users, it is good to divide the individual users into related groups and assign them weight corresponding to their significance for the given activity or a group of activities. For example, for the task of controlling an action of fire rescue units on a large space, a group of operation control will get a bigger weight than the individual units in action. On the contrary, for the task of solution of an action at a concrete place, a group of controlling of the action will get bigger weight than the operation group. It is also good to have information from the individual groups of users which objects and phenomena modelled in the database are more significant for them and which of them are less significant.

The significance of users and the significance of objects may be determined once again based on the user survey with the same methods, just like with the evaluation of significance of products and criteria. In another user survey it is possible to find out in the given groups of users the level of satisfaction or dissatisfaction with the individual objects or their logical groups (e.g. satisfaction with the declared model of communication networks, or declared model of 1st class roads).

Assuming that there are n groups of users where each group has the weight v_u, and m objects and phenomena or their logical groups with weight v_o, based on the conducted user survey it is possible for each object or phenomenon or their logical group to set the level of dissatisfaction β_j:

$$\beta_j = \frac{\sum_{i=1}^{n} \beta_{ij} v_{u,i}}{\sum_{i=1}^{n} v_{u,i}}, \tag{4.11}$$

where β_{ij} is the level of dissatisfaction of i-group of users with j-object.

The resulting value of the coefficient of the complete dissatisfaction α_{11} is then given by relation:

$$\alpha_{11} = \frac{\sum_{j=1}^{m} \beta_j v_{o,j}}{\sum_{j=1}^{m} v_{o,j}}. \tag{4.12}$$

Another group of criteria is created by criteria of *maintenance of the required level of detail of data*. The level of detail is understood as the smallest recognizable detail that can be calculated with. As for solution of their application tasks a user requires corresponding level of detail in the geometric as well as in attributive part, it is necessary to further divide this criterion into two partial criteria of 2nd group—*maintenance of geometric level of detail* and *maintenance of thematic level of detail*. Lowering the required level of detail leads to the loss of accuracy of results of application task solutions, an increase, on the contrary, leads to a growth of expenses for creation and maintenance of the database, furthermore to an increase of requirements on computing, an increase of time for solving application tasks, and last but not least very often also to an increase of complexity of the solution.

Both criteria may be expressed as a percentage share of the suitable objects and phenomena out of all modelled objects and phenomena defined in the given database:

$$k_{121} = 100 \frac{n_{121}}{n_d}, \tag{4.13}$$

$$k_{122} = 100 \frac{n_{122}}{n_d}, \tag{4.14}$$

where

- n_d is the number of all objects and phenomena defined in the database,
- n_{121} is the number of objects and phenomena in the database which in terms of geometric level of detail comply with user's requirements,
- n_{122} is the number of objects and phenomena in the database which in terms of thematic level of detail comply with user's requirements.

The value of criterion *maintenance of the required level of detail of data* may be expressed by a simple aggregation function with consideration of weights of the individual auxiliary criteria:

$$k_{12} = \frac{\sum_{i=1}^{2} p_{12i} k_{12i}}{\sum_{i=1}^{2} p_{12i}}. \tag{4.15}$$

The value of the complete criterion *content of database* may be expressed by formula:

$$k_1 = \frac{p_{11} k_{11} + \sum_{i=1}^{2} p_{12i} k_{12i}}{p_{11} + \sum_{i=1}^{2} p_{12i}}. \tag{4.16}$$

Technical Quality of Database Content

Technical quality of database content and data saved in it is an important criterion which significantly influences the utility value and the quality of digital geo-information. As opposed to the previously mentioned criterion, here based on objective reasons it is necessary to consider different quality of the individual parts of the database. This is caused first of all by the fact that from production-technological and organizational reasons there is never the situation when the complete database has absolutely the same characteristics. At least certain parts will always be less up-to-date, they will have less filled elements, etc. From the practical point of view, it is good to consider the value of the stated criterion in relation to units of measure (UM), in which the database is created, and for which the production capacity and expenses are calculated. The size of these units is usually given by their dimensions stated in geographic coordinates or in planar coordinates or by combination of both. However, if there are products from smaller territorial units being created, or with a lower content richness and small requirements on up-to-dateness, it is possible to evaluate the technical quality of the product and its database in total, without its division to units of measure.

Again, this criterion is not simple, it has its components. Partial criteria of 1st group include the evaluation of source Materials and methods of their transformation, also the positional and attribute data accuracy, their logical consistency and complexity of saved objects and phenomena. Due to methodical reasons, some of the criteria are further divided into partial criteria of 2nd group. To evaluate the complete quality of database means to evaluate all of its components. However, first it is necessary to precisely define these components and then to set the way of their quantification or measurement. The structure of the complete criterion is shown in Table 4.7 which is a part of Table 2.2.

The first part of the stated partial criterion is *the transparency of origin of source material when collecting primary data.* Here it is necessary to evaluate whether the database creators know the origin and characteristics of all source materials that are used when filling it. Provided that the database creators know exactly the given characteristics, the value of this criterion equals 100. If the characteristics are not exactly known, the value is lowered by the percentage share of unknown or incomplete information expressed by number α_{211}. Then for x-unit of measure it is possible to calculate:

$$k_{211}^x = 100 - \alpha_{211}^x. \tag{4.17}$$

Another part of the partial criterion is *technical correctness of the used methods and models for secondary data derivation.* The used methods and mathematical models when deriving secondary data may significantly influence the resulting accuracy of such data (see e.g. Longley, Goodchild, Maguire, & Rhind, 2016). Similarly to the previous criterion, its value equals 100 in case the database creator and administrator provides complete information. If the information about the used methods and models are not exactly known, the value is decreased by percentage share of the unknown or incomplete information expressed by number α_{212}. So it is possible to calculate:

$$k_{212}^x = 100 - \alpha_{212}^x. \tag{4.18}$$

Table 4.7 Structure of criterion *technical quality of database content*

Group of main characteristics—main criteria	Characteristics of partial criteria	Definition	Quality criterion
Technical quality of database content—k_2	Transparency of data sources and methods for secondary data derivation—k_{21}	Transparency of source materials on primary data collection	Level of availability of information about used sources—k_{211}
		Transparency of methods and models used for secondary data derivation	Level of availability of information about used methods—k_{212}
	Positional accuracy—k_{22}	Compliance with declared horizontal accuracy	Percentage of objects with satisfactory horizontal accuracy—k_{221}
		Compliance with declared elevation accuracy	Percentage of objects with satisfactory elevation accuracy—k_{222}
	Attribute accuracy—k_{23}	Compliance with declared accuracy of attribute data	Percentage of objects with satisfactory attribute accuracy
	Logical consistency of database—k_{24}	Degree of adherence of geographic data (data structure, features, attributes, and relationships) to the models and schemas (conceptual model, conceptual schema, application schema, and data model)	Percentage of objects with topological consistency—k_{241}
			Percentage of objects with attribute consistency—k_{242}
			Percentage of objects with time consistency—k_{243}
	Data completeness—k_{25}	Degree of adherence of the entirety of geographic data (features, attributes, and relationships) to the entirety of the modelled part of landscape	Percentage of missing objects of surplus objects—k_{251}
			Percentage of properties of thematic objects that are complete—k_{252}

The complete value of partial criterion k_{21}^x is then given by relation:

$$k_{21}^x = \frac{\sum_{i=1}^2 p_{21i} k_{21i}^x}{\sum_{i=1}^2 p_{21i}}. \tag{4.19}$$

Second partial criterion—*positional accuracy*—evaluates the accuracy of placing objects and phenomena in the given geodetic reference coordinate and elevation system. The evaluation is based on a statistical comparison of chosen samples of the database with primary sources or with standards of higher accuracy from which the expected values are then calculated.

The positional accuracy is strongly bound to geometric *level of detail of data*. This expression with digital data substitutes the expression *scale* used with graphical products. Geometrical level of detail indicates with which maximum accuracy the data are provided, or in other words how big their smallest recognizable detail will be, or what the minimal spatial distance between data elements is (STANAG 2215, 2010). The positional accuracy itself then expresses the real level of reached accuracy of localization of objects and phenomena and their characteristics. The characteristics may be directly measured by objective procedures.

Positional accuracy depends on many influences given by objective or subjective factors.

Among the objective factors it is possible to include:

– Influence of distortion of the used cartographic projection,
– Influence of accuracy of positional and elevation geodetic basis,
– Influence of technical realization of cartographic sources in case these are considered to be the default basis for filling the database,
– Influence of limits of technological procedures during field collection of data in case these are used for filling the database,
– Influence of limits of technological procedures when using contactless methods of data acquisition (e.g. orthorectification of aerial photos or data from distant research of the Earth, etc.) in case these are used for filling the database.

All objective influences may be precisely characterized by size as well as by localization in the given geodetic reference system. Although objective influences cannot be removed, they can be used for a calculation and eliminated by suitable computing methods when using the data. Moreover, the producer of data should always choose such sources and such procedures that guarantee corresponding data accuracy while keeping the given level of detail. Especially elimination of serious and systematic mistakes should be guaranteed.

Serious mistakes and most of systematic errors should be removed in the process of the whole technological cycle of production or updating of databases by thorough and systematic checks. However, it is necessary to accept random mistakes and consider their occurrence when using the data. The data provider should provide accurate information about their occurrence. According to STANAG 2215 (2010), paper as well as digital products are evaluated by categories in which a horizontal—also vertical—accuracy of the resulting product is set (Table 4.8).

Table 4.8 Absolute accuracy in geometry within GRS WGS84. CMAS is the Circular Map Accuracy Standard and LMAS is the Linear Map Accuracy Standard (STANAG 2215, 2010)

Rating	Horizontal and vertical ratings in terms of WGS84 datum	
	CMAS rating	LMAS rating
A	A	0
B	B	1
C	C	2
D	D	3
E	Product not referenced to WGS84	

Horizontal accuracy is defined by the position of satisfyingly defined elements, on which their position with reference points is tested. The position of these points is set by independent methods and it is taken as correct. In order for the product to be put in the appropriate category, the number of satisfyingly defined points must comply in their position with the set limits given in Table 4.9. In terms of statistics, the level of reliability 90% is chosen here.

The producer of data in the sense of the previous paragraph puts the product into the appropriate category. An independent testing of horizontal accuracy according to the same methodology may prove legitimacy or illegitimacy of the classification into the appropriate category. Criterion k_{221} then from this point of view may evaluate functionality of x-unit of measure of the product. Its value may be calculated:

$$k_{221}^x = 100 \frac{n_{221}^x}{n^x} + h_s, \tag{4.20}$$

where

- n^x is the number of all objects and phenomena in x-unit of measure of the database,
- n_{221}^x is the number of objects and phenomena in x-unit of measure of the database which comply with the given category in horizontal accuracy,
- h_s is a chosen level of reliability given in percentage.

Table 4.9 Absolute horizontal accuracy (STANAG 2215, 2010)

Category (rating)	Measurement at product scale (mm)	Map scale or corresponding resolution of digital data (m)				
		1:25,000	1:50,000	1:100,000	1:200,000	1:250,000
A	0.5	12.5	25	50	100	125
B	1.0	25	50	100	200	250
C	2.0	50	100	200	400	500
D	>2.0	Worse than category C				
E	Not set					

Elevation accuracy is evaluated similarly. Products may be again categorized according to the reached and tested elevation accuracy on reference points. Categorization is defined by standards, e.g. STANAG 2215, (2010). According to this standard, the product is put into an appropriate category based on finding differences of elevation of the evaluated product and standard of quality. To make sure the product is put into an appropriate category in Table 4.10, it is necessary that 90% of control points meet the limits of the given category.

Independent testing of elevation accuracy according to the same methodology may again prove legitimacy or illegitimacy of the classification into an appropriate category. From this point of view, using criterion k_{222} then may evaluate the functionality of the product in the space of x-unit of measure. Its value may be calculated:

$$k_{222}^x = 100 \frac{n_{222}^x}{n^x} + h_s, \tag{4.21}$$

where

- n^x is the number of all objects and phenomena in x-unit of measure of the database,
- n_{222}^x is the number of objects and phenomena in a unit of measure in database which in terms of horizontal accuracy comply with the given category,
- h_s is a chosen level of reliability given in percentage.

The resulting value of the partial criterion k_{22}^x is then possible to calculate according to relation:

$$k_{22}^x = \frac{\sum_{i=1}^{2} p_{22i} k_{22i}^x}{\sum_{i=1}^{2} p_{22i}}. \tag{4.22}$$

The third partial criterion is *an attribute accuracy*. Also here, however, it is necessary to distinguish expressions *accuracy* and *level of detail of thematic data*. The required level of detail of thematic data is set beforehand by the technical project of terrain model and user is informed about a number and size of the smallest recognizable thematic detail. From the point of view of the user, its value influences the usability of the model towards the solution of application tasks, or the resulting

Table 4.10 Absolute vertical accuracy of digital products

Category	Map scale or corresponding resolution of digital data				
	1:25,000	1:50,000	1:100,000	1:200,000	1:250,000
0	2.5 m	5 m	10 m	20 m	25 m
1	5 m	10 m	20 m	40 m	50 m
2	10 m	20 m	40 m	80 m	100 m
3	Worse than category 2				
4	Not set				

accuracy of these applications. Attribute data accuracy is a criterion that is used to evaluate the quality of technical realization of the product.

Every data producer or provider should evaluate the attribute accuracy and inform the user about the results of the evaluation, e.g. by rating the products into a certain classification class, i.e. similarly to the case of positional accuracy.

In order to establish the real attribute accuracy, it is good to test it also independently. For evaluation—just like with positional accuracy—comparison standards or test data are used. With regards to the level of detail of the evaluated product, they may be considered flawless. Attribute accuracy of quantitative (measurable) characteristics is most often evaluated by a mean error, with qualitative characteristics by various classification procedures (with discreet objects) or an error matrix (for continuous phenomena).

It is possible to evaluate the functionality of a product based on the results of the mentioned independent tests using criterion k_{23}^x as a percentage share of objects and phenomena with correct (in the given classification class) thematic attributes against all objects and phenomena in the database. From the point of view of error distribution, the level of reliability h_s applies. It is possible to write:

$$k_{23}^x = 100 \frac{n_{23}^x}{n^x} + h_s, \tag{4.23}$$

where

- n^x is the number of all objects and phenomena in x-unit of measure of the database,
- n_{23}^x is the number of objects and phenomena in x-unit of measure of the database which are in terms of attribute accuracy compliant with the given category,
- h_s is a chosen level of reliability given in percentage.

The fourth partial criterion is *logical consistency of database*. Logical consistency means that there are all needed assignments between the individual objects and phenomena and their characteristics are realized and permanently maintained in the database, which enables to analyze relations and connections between these objects. This criterion evaluates the existence of obvious discrepancies in the whole database that may occur during its initial filling as well as during updating. It evaluates *topological consistency* as a basic presumption for the function of most application tasks, also *attribute consistency* and *time consistency*.

It only has sense to speak about *topological consistency* with databases whose topology is expressed, i.e. first of all vector databases. Each topological level is defined by:

– a list of mandatory and optional topological elements,
– a list of mandatory and optional topological relations,
– a list of topological rules and restrictions that involve the existing set of topological elements.

Breaking the rules of the given topological level may lead to breaking the topological consistency when using the database and especially during its updating. Inconsistencies may be discovered by testing the database with independent tests which for example for level 2 evaluate:

– Disconnection of the node to the edge,
– Missing relation of an edge to an object,
– Non-alignment of objects,
– Error in initial coordinates of an edge,
– Error in number of an edge,
– Zero length of an edge,
– Wrong number of nodes,
– Wrong number of edges,
– Wrong length of block of nodes,
– Wrong length of block of edges.

The level of security of topological consistency may be expressed by criterion k_{241}^{x} as a percentage number of topologically consistent objects of all objects in x-unit of measure of the database, i.e.:

$$k_{241}^{x} = 100\frac{n_{241}^{x}}{n^{x}}, \qquad (4.24)$$

where

– n^{x} is the number of all objects and phenomena in x-unit of measure of the database,
– n_{241}^{x} is the number of objects and phenomena in x-unit of measure which are consistent in terms of topology.

Attribute consistency means that there are no redundant occurrences of attribute domains in the database which have the same name, yet not containing the same data. In databases (especially relational) it is very important that also this attribute consistency is permanently maintained, however it is necessary to point out that there will always be purpose redundancies in extensive databases. However, these redundancies must be taken care of by the system of database control beforehand.

One of the ways to maintain attribute consistency during designing the data model is to perform *normalization*. The aim of normalization is (Kresse & Danko, 2012):

– to enable representation of each relations in the database,
– to get efficient searching algorithms based on a simpler set of relation operations than it would be possible otherwise,
– to remove unwanted dependencies in relations during operations of filling, updating and canceling,

- to reduce the need of restructuring while implementing a new type of data,
- to make sure that the summary of relations is neutral in regards to the frequency of inquires that tend to change in time.

The process of normalization lies in a gradual standardization of the database into so-called informal forms. After the process of normalization the database should be in a thematically consistent state and resistant towards degradation of functionality while being updated.

Attribute consistency may be tested by independent tests, with the help of which occurrences of redundant data and dysfunctional and transitive dependencies are discovered. The level of security of attribute consistency may then be expressed by criterion k_{242}^x as a percentage number of thematically consistent objects of all objects in x-unit of measure of the database, i.e.:

$$k_{242}^x = 100 \frac{n_{242}^x}{n^x}, \qquad (4.25)$$

where

- n^x is the number of all objects and phenomena in x-unit of measure of the database,
- n_{242}^x is the number of objects and phenomena in x-unit of measure which are consistent from the thematic point of view.

The last part of logical consistency is *time consistency* which relates to data in domains for storage of time data. If the database is in time consistent state, an occurrence of two different events in one time is not possible. Breaking this rule leads to occurrence of inconsistencies where in one place there might wrongfully appear two or even more events happening at the same time. The result of this is that the user will not know which phenomenon in the given time happened or is just happening. That is the reason why it is necessary to test also this time consistency. Its level after conducting independent tests may be expressed by criterion k_{243}^x as:

$$k_{243}^x = 100 \frac{n_{243}^x}{n^x} \qquad (4.26)$$

where

- n^x is the number of all objects and phenomena in x-unit of measure of the database,
- n_{243}^x is the number of objects and phenomena in x-unit of measure of the database that are time consistent.

The result value of partial criterion k_{24}^x is calculation the same way as the previous partial criteria:

$$k_{24}^x = \frac{\sum_{i=1}^{3} p_{24i} k_{24i}^x}{\sum_{i=1}^{3} p_{24i}}. \qquad (4.27)$$

The last partial criterion is *data completeness*. Data completeness evaluates the completeness of fulfilling of all prescribed objects and phenomena and their characteristics. However, many objects may be included in the database without thematic attributes having been given. The reason may be restricted capacities of the data producer or inability to get relevant or trustworthy thematic attributes. That is why the user encounters this fact quite often. From practical reasons it is good to evaluate separately the *complexity* of fulfillment of the individual *objects and phenomena* and the *complexity* of fulfillment of their *thematic attributes*.

A higher number of objects in the database than in reality or than it is declared in the database description is a rare—but not excluded—phenomenon. The reason may be an incorrect interpretation of the description of the data representation by a concrete worker, or the use of a catalogue of objects for another product than the product that is being worked on. Although such situations should be taken care of by the system of running checks and final inspection when the product is created, it is not possible to fully exclude them. When using product database in applications however, also excess number of objects may lead to faulty results of analyses made with such data.

Both criteria are evaluated in percentage and that's why it is possible to set evaluation criteria k_{251}^x and k_{252}^x as a percentage share of technological directives of prescribed objects and phenomena really filled in the database, or prescribed and really filled thematic attributes. Thus:

$$k_{251}^x = 100\frac{n_{251}^x}{n^x}, \qquad k_{252}^x = 100\frac{n_{252}^x}{n^x}, \qquad (4.28)$$

where

- n^x is the prescribed number of all objects and phenomena in x-unit of measure of the database from the given locality,
- n_{251}^x is a number of really fulfilled objects and phenomena in x-unit of measure of the database,
- n_{252}^x is a number of objects and phenomena in x-unit of measure of the database which have all prescribed thematic attributes fully completed.

The complete value of partial criterion k_{25}^x is calculated in the same way as with the previous partial criteria:

$$k_{25}^x = \frac{\sum_{i=1}^{2} p_{25i} k_{25i}^x}{\sum_{i=1}^{2} p_{25i}}. \qquad (4.29)$$

The complete value of the criteria evaluating the quality in x-unit of measure of database k_2^x is possible to calculate according to relation:

$$k_2^x = \frac{\sum_{i=1}^{5} p_{2i} k_{2i}^x}{\sum_{i=1}^{5} p_{2i}}. \qquad (4.30)$$

Weights of the individual partial criteria of 1st group and 2nd group can be set based on expert surveys (Talhofer, 2002).

Database Timeliness

Database Timeliness expresses that the whole database of the evaluated product is in the up-to-date state or that its most significant parts are in this state. Since the whole system of quality evaluation relates to products which contain stored relatively stable objects, the timeliness in relation to on-line data will not be discussed here as their time consistency is usually limited to the duration of the real occurrence of an object or phenomenon. Thus, database objects modelling phenomena such as traffic event, immediate vehicle position, etc. will not be evaluated.

Timeliness of a database or its parts (object groups) is understood first of all as a level of compliance of its content with the modelled reality in the given time, i.e. the level of its obsolescence. Generally speaking, each database is obsolete to certain extent already at the moment of the filling. From production-technological reasons it is basically not possible to reach a state when the database content is fully compliant with the reality. As opposed to classic paper maps, database may be updated incomparably more often, in many cases also in real-time. However, this will concern most of all the most significant objects and phenomena, the basic part of the database will be in the state in which the provider delivers it.

The level of database timeliness changes quite quickly depending on various factors. Its value may be expressed by percentage number of changes that are made in geometry, as well as in topology or objects attributes. In meta-information files of corresponding databases, or their parts, it is possible to observe and evaluate these changes permanently. Yet, it seems purposeful to evaluate the level of timeliness as an aggregate function dependent on time that passed since the last database update. In principle, this can be done similarly as with classic paper maps. Thus, it is possible to evaluate not only timeliness of the individual elements of their logical groups, but also homogeneity of timeliness of the entire database and the system of its maintenance. When evaluating timeliness of the individual elements or their logical groups it is possible to take into account also their significance for a solution of type tasks in decision-making processes.

> Note: It is necessary to distinguish between *time accuracy* and *timeliness*. *Time accuracy* informs the user about the shortest recognizable detail in the determination of time of saving or change of an object or phenomenon and their attributes. It is usually expressed in time units. It is set beforehand in a technical project of terrain model. On the contrary, *timeliness of database* expresses the level of its obsolescence against the real situation in terrain in a given moment of time. However, at the same time it is a specific tool for measurement of time accuracy.

Due to human activities, the countryside changes quite quickly. In cases when a critical situation is being solved, e.g. a war conflict or natural disaster, we may expect that changes on the countryside will be even more significant than in everyday conditions. Due to ongoing changes, the content of databases deteriorates and thus also suffers loss of their utility characteristics.

From the point of view of the data model, the changes are first of all:

– The creation of new objects and phenomena of the evaluated type,
– Extinction of existing objects and phenomena of the evaluated type,

– Change of geometry of existing objects and phenomena of the evaluated type and thus also a change of their topological characteristics,
– Change of quantitative characteristics of the current objects and phenomena of the evaluated type,
– Change of qualitative characteristics of the current objects and phenomena of the evaluated type,
– Change of name of the current objects and phenomena of the evaluated type.

From their course of time, changes of objects and phenomena may be divided to:

– relatively regular in predefined dimensions,
– periodically variable mostly in the daily or yearly period,
– irregular in pre-definable dimensions,
– irregular in pre-undefinable dimensions.

The mentioned division has its significance not only for the evaluation of timeliness of a database, but also from the point of view of the possibility to model these changes mathematically beforehand and thus increase the timeliness of a database.

Real changes in the database content may be found out only by a comparison of its content and the modelled reality, e.g. by direct observation of terrain, by a comparison of the database content with aerial or space photos, with a permanently maintained database of changes, with registration maps of changes, etc. With regards to the character of observation of the utility value of the entire database, to express its timeliness, it is possible to use also the time difference between the time when the database is evaluated and the time when its content complied with reality. It is possible to proceed from the following general principles (Miklošík, 1987):

– The complete range of changes, i.e. level of obsolescence, on the observed area increases with time that has passed since the last update of database content. However, this increase of the complete obsolescence of content in a time unit tends to decrease.
– The speed of changes of the individual objects and their logical groups is—in general—various. Altimetry belongs to the most stable, settlements change the most.
– Areas with high density of objects and phenomena of social-economic sphere (city agglomerations, intensive mining activity, …) change much faster than areas of mostly agrarian character with low density of population.
– The pace of obsolescence is the quicker, the higher geometric and thematic resolution of objects is. On the contrary, with a decreasing level of detail the pace of obsolescence is slowing down.

The entire compliance of the database content and the modelled reality is in time T_0. However, it is not possible to reach such an ideal state because there is always a time delay between the initiation of working on filling data and their provision to users. That is why the initial point for further evaluation is time T_{ed}, for which the set compliance of the modelled reality and database content applies. It is usually stated in regulations for its creation.

If the database content is not updated, it will gradually lose its timeliness and eventually it will reach a certain level (in time T_{lim}), when its usage will be a problem, i.e. it will lose the ability to serve the purpose for which it was built. This limit value of obsolescence was quite often seen with an evaluation of classic maps where there is loss of utility already in cases when there is 15–25% of the map content changed, (Miklošík & Rybansky, 1997). It is caused also by the fact that each discrepancy in the map content casts doubt on users also with other elements, even though these may be correct. The size of the mentioned influence, however, depends also on user's experience because an experienced user is able to work very well also with a quite obsolete map.

For digital data this question has not been properly discussed yet, nor corresponding results have been published. However, it is necessary to realize that we have not been working with digital data for the same time as with classic maps, and that their use is more extent and in many cases these data enter into various computing systems (navigation, control, …) directly in the digital form where it is not possible to expect any user's "experience". Faulty input data may lead to wrong results and often even to very serious consequences due to decisions based on these results. Here it is essential to evaluate the timeliness of the database in relation to user functions and evaluate this criterion according to the prevailing function. If an information function of the used database is the prevailing function, then it is possible to proceed from the experience with work with paper maps. In case the database has the prevailing usage as a means of automatization or as a source for mathematical modelling, it is necessary to expect its high sensitivity to its timeliness (Talhofer, 2004; Talhofer & Hofmann, 2009; Talhofer, Hoskova-Mayerova, & Hofmann, 2012).

It can be assumed from these considerations that a digital database basically may completely lose its utility value when reaching limit T_{lim}. In general, we may say that the pace of obsolescence of digital geo-information will be higher than with classic paper maps. However, it will also depend on the application program that uses DGI.

A change of timeliness of the database content may be expressed by $f(T)$ which models relations of such change and with the help of which it is possible to predict the development of the utility value of the database depending on the time that has passed since it was last updated. Using Miklošík's theory given in (Miklošík, 1987) it is possible to construct it based on various presumptions.

The solution may be a presumption that the change of timeliness of content occurs with linear dependence on time. Then the sought after function will be in the form:

$$f_1(T) = \frac{(1 - C)(T_{lim} - T)}{T_{lim} - T_{ed}} + C = \frac{C(T - T_{ed}) - (T - T_{lim})}{T_{lim} - T_{ed}}. \tag{4.31}$$

Constant $C \in <0, 1>$ and expresses the limit value of database functionality in terms of time. Basically it is given by the ratio of very stable objects of the modelled reality on the total number of objects. Since from the logic of previous considerations it follows that it has sense to investigate database functionality for time $T \leq T_{lim}$, then for $T > T_{lim}$, it is possible to write:

$$f_1(T) = C.$$

Provided that the size of a decrease of the level of timeliness of the database content in a time unit shrinks depending on the increasing time T, it is possible to use an exponential form of the function:

$$f_2(T) = e^{-b(T-T_{ed})}, \tag{4.32}$$

where b marks the constant dependent on the limit time of obsolescence which might be expressed by relation:

$$b = \frac{d}{T_{lim}}. \tag{4.33}$$

Constant d is chosen from interval $<1, 2>$ depending on the relative representation of objects and phenomena with various time stability. In an ideal case when objects and phenomena saved in the database all change in the same pace; the size of constant d equals 1.

In the most unfavourable case when the pace of change of objects and phenomena would be rather disparate, the constant equals 2. Based on cartometric research carried out in the 80s of 20th century on the territory of the Czech Republic it was found out that changes on our territory are rather inhomogeneous, and thus basically with all map sheets the value of constant d is close to two (Miklošík, 1987). Due to the uneven development of the territory also in the following years it is possible to fully adopt this value into the evaluation of timeliness of digital products as well.

The real size of this constant for databases, however, must be experimentally verified. The graph (Fig. 4.3) shows a modelled influence of value d on the value of function $f_2(T)$.

A significant reason which with increasing time T decreases the pace of decrease of timeliness of the database content is the fact that once changed elements of its content cannot manifest in worsening of its compliance with reality in the next period. Based on this consideration it is possible to write the function of content timeliness also in the form (Miklošík & Rybansky, 1997):

$$f_3(T) = \left[1 - \frac{1}{T_{lim}}\right]^{T-T_{ed}}. \tag{4.34}$$

Assuming that a part of the database content in the country is so stable that it may be with a certain proximity considered "timeless" (e.g. elevation model or altimetry), the function for the expression of a change of timeliness of the database content will be in the form:

$$f_4(T) = (1 - C)\left[1 - \frac{1}{T_{lim}}\right]^{T-T_{ed}} + C, \tag{4.35}$$

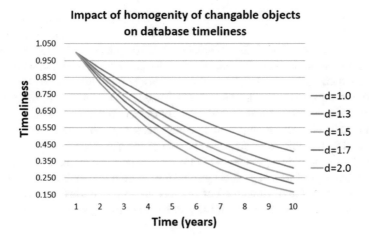

Fig. 4.3   Impact of homogeneity of changeable objects on database timeliness

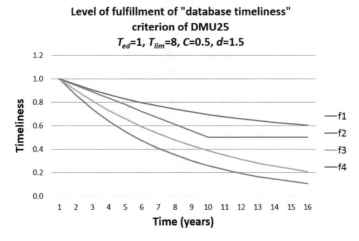

Fig. 4.4   Graphs of functions of progression of database obsolescence

where C marks a relative share of very stable elements on the complete database content.

The graph (Fig. 4.4) shows the graphs of the individual functions with the state initial parameters.

Functional relations stated in the previous paragraph enable to predict the level of timeliness of the whole database in the given locality. The significance of this way of evaluation lies especially in the fact, that it is possible beforehand and with certain probability to set how the database will behave with regards to time period since its last update, i.e. how obsolete in a certain moment it will be. The mentioned function $f_4(T)$ may be used for the entire database as well as for their parts, e.g. to units of measure, for a given homogenized territory (city and industrial agglomerations, agrarian areas) or for chosen layers. That is why it is possible to use these functions for instance for planning of the complete database update or also for the decision which parts of the database are necessary to be quickly updated and which are so stable that their update for the given purpose is not absolutely necessary yet. Thus, it is good to evaluate also the database timeliness with consideration of differentiated timeliness for each individual logical group of objects and phenomena.

With the theoretical findings from the previous paragraph and based on statistical research, it is possible to evaluate the course of obsolescence of objects and phenomena in the individual groups. Furthermore, it is possible to find out when for the individual groups of users a j-object group stops to be functional, i.e. what $T_{lim,j}$ is. To evaluate the value of timeliness of j-object group it is basically possible to use one of functions f_1, f_2, f_3, or f_4. If for instance function f_2 is used, it is possible to write:

$$f_{2,j}(T) = e^{-b_j(T-T_{ed,j})}, \tag{4.36}$$

where

$$b_j = \frac{d}{T_{lim,j}}.$$

The value of constant d again expresses relative representation of objects and phenomena in the given groups with various time stability.

If it was necessary to express the value of timeliness function of the entire database with consideration of differential timeliness for the individual object groups, then it would be possible to calculate it as a weighted average where $v_{o,j}$ is the weight of j-object group:

$$f_2(T) = \frac{\sum_{j=1}^{m} v_{o,j} f_{2,j}(T)}{\sum_{j=1}^{m} v_{o,j}}, \tag{4.37}$$

where m is the number of object groups.

Area Importance

Criterion *area importance* k_4 is given by the users' needs so that it meets their requirements on a spatial range of the area being processed or secured. The significance of this criterion is given especially by the fact that for each user the territory is of a different significance, e.g. military, political, economic, etc. However, databases containing data from different territories are not mutually replaceable in any way. Here it is essential to consider that in the given space each part is not of the same significance for every user. For instance, for defense of a country against migration waves of large extent, its border zone is of higher significance than inland agglomerations; areas around larger water courses are significant for evacuation of inhabitants in case of massive floods; areas around large chemical plants are of high significance for dealing with potential ecological disasters, etc. Furthermore, it is necessary to consider that absence of data from certain spaces may lead to restricted functionality of the entire database.

Criteria for evaluation of area importance express such characteristics of territory and phenomena happening on them (past, present, as well as future) which relate to the territory and which directly or indirectly raise interest in the given territory. That is the reason why the first and essential step for making a suggestion of evaluation criteria and their weights is to make research of users' needs and requirements and summary of findings about who, what for and in what extent, or how often uses or will use the certain type of database, or which of its parts in terms of territory.

Setting of concrete criteria depends first of all on whether and to what extent the range of users and the purpose and way of usage of certain digital geo-information is defined beforehand. In terms of DGI used in the army, with high probability it is possible to set the following structure of partial criteria:

1. Geographic position of the evaluated part of territory,
2. Entering corridors into the part of interest of territory,
3. Number and character of obstacles,
4. Centers of industrial production,
5. Population density,
6. Regions with concentration of national or religious minorities,
7. Places of deployment of garrisons and their size,
8. Military training areas,
9. Warehouses of emergency, operation and other supplies,
10. Allocation of systems and equipment securing territory defense.

The stated criteria certainly are not the complete list, yet they may be used to document the range of problems that are necessary to consider while evaluating the area importance. They may be gradually added, reduced, grouped, etc. However, from the point of methodology of work it is possible to work with the given criteria with sufficient objectivity. For their evaluation it is possible to derive physical-geographic criteria (geographic position, entering corridors, natural obstacles, etc.), social-economic criteria (range of production, population density, density and configuration of infrastructure network, regions of religious or national minorities, …) and military-political criteria from them (deployment of garrisons and their size, military training areas, warehouses of emergency, operation and other supplies, allocation of systems and equipment providing defense of the country). All of these criteria are related to stable or relatively stable objects saved in databases. In terms of using DGI in crisis management, it is necessary to complement these criteria with criteria relating to unstable objects, such as:

1. Deployment of intervention units in the territory,
2. Key terrain for solution of a current situation set e.g. based on an analysis of extent and seriousness of critical situations (accident of a large extent, territories hit by floods, etc.).

Table 4.11 Weights of partial criteria of criterion area importance

Main criterion	Partial criterion of 1st group	Weight of criterion
Area importance—k_4	Geographic position of the evaluated part of territory—k_{41}	0.143
	Entering corridors into the part of interest of territory—k_{42}	0.096
	Number and character of obstacles—k_{43}	0.101
	Centers of industrial production—k_{44}	0.113
	Population density—k_{45}	0.091
	Regions with concentration of national or religious minorities—k_{46}	0.081
	Places of deployment of garrisons and their size—k_{47}	0.079
	Military training areas—k_{48}	0.088
	Warehouses of emergency, operation and other supplies—k_{49}	0.114
	Allocation of systems and equipment securing territory defense—k_{4_10}	0.094
Total		1.000

Note With regards to the fact that there are more than 9 partial criteria, for formal reasons the 10th criterion is marked as k_{4_10}. In case other partial criteria are used, it is possible to proceed similarly

Each criterion has its own weight. It is usually set based on users' research from crisis management, e.g. by the method of pair comparison or the method of direct estimation in the given classification scale. The Table 4.11 shows the above-mentioned criteria provided by users' research among workers of Geographic service of the Czech Army.

The peculiarity of the weighting of the individual criteria is the fact that some of them are variable in time. While geographic position, entering corridors, etc. are basically invariable in time (e.g. corridor of Moravian Gate in the Czech Republic), social-economic criteria and social-political criteria depend on the given political situation and thus they are variable in time. This situation may be either quite stable in time, or it can change very often, or it can also be restricted in time.

Before the criterion evaluation, it is necessary to define the area of interest (AOI) in which the activity will take place. In AOI objects and phenomena that are important for the solution of the task are chosen from the database. It is also possible to require the database administration to add objects and phenomena that are temporary or missing in the database. The importance of chosen objects and phenomena in AOI will be maximal in the place of their occurrence and it will decrease with an increasing distance from them. With the help of analytical tools GIS it is possible for every place in the given space to set an influence of a concrete object or a group of objects divided according to the structure of partial criteria of 1st group. One of the tools is a calculation of Euclidean distance or the use of Inverse Weight Distance tool (IDW). While when calculating Euclidean distance weights of objects modelling their significance are not considered, IDW method may apply these weights (Esri, 2013). The result of the analysis is a raster data file in which the value of raster cell corresponds to the value of partial criterion of 1st group. The resulting value of criterion k_4 for the given raster cell may be calculated with the help of tools of map algebra.

For evaluation of area importance according to the individual criteria it is possible to choose for instance the following methodology:

1. Division of the area of interest (given area) to squares, usually raster cells (units of measure).
2. The value of partial i-criterion in x-unit of measure is inversely proportional to its distance from the closest object evaluated according to i-criterion (e.g. entering corridor to an area, distribution of warehouses, etc.).
3. Appropriate analytical tools of GIS (Euclidean distance, IDW, density function) are used to set the value of partial criterion. Their results are re-classified into a suitable classification scale, most often $0,\ldots,10$.
4. To set the total level of evaluation of area importance based on carried out partial criteria, it is possible to use a multiplication function in the form:

$$k_4^x = \sum_{i=1}^{n} p_i k_{4i}^x,$$ (4.38)

where

- k_4^x is the total evaluation of x-unit of measure,
- k_{4i}^x is a partial evaluation of x-unit of measure according to i-criterion,
- p_i is the weight of i-partial criterion,
- n is the total number of used partial criteria.

The multiplication function is chosen with regard to the specific meaning of the first criterion, the geographic location of evaluated area. This criterion is also used to evaluate the fact that the producer is not able to give relevant data from the user's area of interest, or he/she is able to provide data only from a part of AOI.

The whole area importance may then be expressed as an average value of criteria of all evaluated units of measure. This value, however, is of approximate significance only; in terms of evaluation of area importance, spatial distribution of places with various value of criterion k_4 is more important.

The methodology of setting the significant coefficient is documented in the following example. Near city Brno, Czech Republic, an area of responsibility (AOR) was defined, which complies with the required AOI. The database from DMU25 was secured from this area (Fig. 4.5).

As in the chosen area there were not represented all types of evaluated objects, or these evaluated objects cannot be published, they were complemented with only modelled ones (Fig. 4.6).

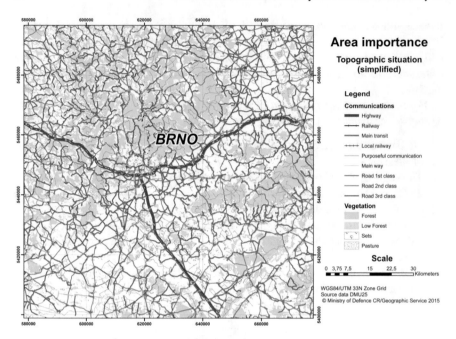

Fig. 4.5 Area of responsibility for the calculation of coefficient of area importance k_4

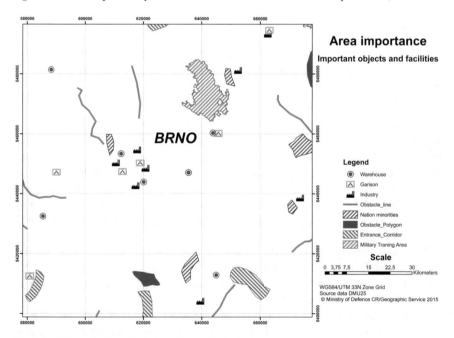

Fig. 4.6 Objects for the calculation of partial criteria

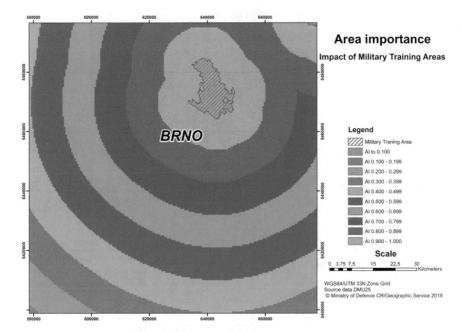

Fig. 4.7 Values of coefficient k_{48} in AOR

With the help of analytical functions, influences of the individual groups of objects were calculated. The result was raster files with the size of a pixel 500 × 500 m (units of measure). The Fig. 4.7 picture shows an example of calculation of value of criterion k_{48} (military training areas) where values of criteria are expressed as values of pixels of a raster field. Function Euclidean Distance was used for the analysis.

Similarly influences of all other object groups were analyzed and they were expressed as values of the individual criteria. The total area importance is calculated for the individual units of measure with the help of tool MapAlgebra using relation of expression (4.38). The result of the analysis can be seen in the picture (Fig. 4.8).

The picture Fig. 4.9 shows spatial distribution of values of coefficient k_4 and there are obvious places where the coefficient reaches high values, as well as places where its values are small. Statistical division of values is shown in the picture (Fig. 4.9), from which it is obvious that the average area importance is valued as $k_4 = 0.514$ (after re-classification). Program system ArcGIS 10.2 was used for all analyses and calculations (Esri, 2013).

A more detailed description of the methodology of spatial analyses is given in Chap. 7.

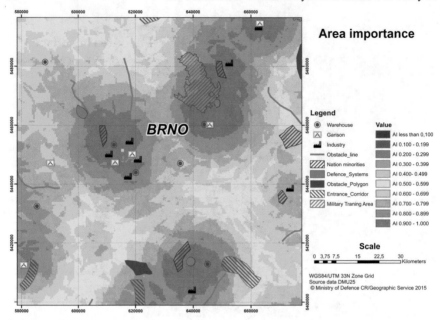

Fig. 4.8 The complete area importance in AOR in values of criteria k_4

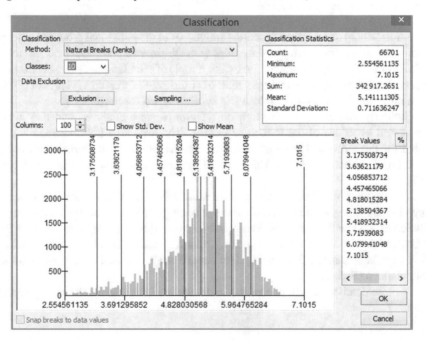

Fig. 4.9 Distribution of values of coefficient k_4 in AOR

User Friendliness

Criterion *user friendliness* is used to evaluate the level of standardization, independence, and data security. It is understood as an evaluation of abilities of data to be used in various types of program environment of GIS character, furthermore as independence of data on a concrete program environment, and finally a system of their security against damage and abuse.

This criterion is also divided into three partial criteria—data standardization, data independence on program environment and data security against damage and abuse.

DGI *standardization* nowadays is a very discussed concept. The principle of standardization lies in the agreement of involved parties that they will provide data in standard exchange formats that will secure their problem-free use in a system accepting the mentioned standards.

Within standardization they are especially the following questions (DGIWG-103, 2008):

- Supported data structures,
- Data formats,
- Concept model of landscape sphere (scheme of coding of objects and phenomena and their attributes),
- Exchange media,
- Administrative procedures.

In the field of geoinformatics there is no worldwide or Europe-wide binding standard of area localized information. The standards also strongly depend on the field, in which digital information is worked with. They will be different in the field of state administration, ecology, forestry and soil management, etc. Organizations that guarantee binding state databases, deal with the questions of DGI standardization within one country, in the Czech Republic it is e.g. State Administration of Land Surveying and Cadastre (CZUK), in the USA it is Federal Geodetic Control Subcommittee and Federal Geographic Data Committee, etc. On the international level it is previously mentioned organizations such as ISO (2017) or Defense Geographic Information Working Group (DGIWG-103, 2008).

However, from the point of view of a user, it is important whether he/she gets the data in a standardized form or not. Of course, it will always be a concrete standard for concrete data that will or will not fully respect the standard. That is the reason why the value of criterion k_{51} is:

- $k_{51} = 0$, if the given standard is not respected,
- $k_{51} = 100$, if the given standard is respected.

From the general point of view, it has no sense to speak about partial standardization. However, if the situation requires it, it is possible to provide partial components of a database in an agreed, nevertheless not fully standardized version. Then it is possible to set the level of standardization as a percentage share of standardized information from all information in the database. Such case may happen especially when solving critical situations when it is necessary to accept e.g. updating data from sources that are not normally used and the securing unit in the occurred situation does not have the possibility to transform these data into standards.

Data independence on program environment expresses first of all their independence so that it is possible without any modifications to use it in various program environments while securing their full usability. In principal, to certain extent is the independence bound to data standardization, yet it seems effective to evaluate it separately.

From the point of view of the producer, the produced data is always bound to the used program equipment. However, the users expect to receive data in such a form that will enable them their problem-free implementation to the device they use. This is again a concrete system equipped with a concrete technology that may be based on GIS technologies or it may be created only as single-purpose.

A lot of command and control systems were created as unique products that accepted DGI in their declared and unchanging standards and any breach of this standardization disables working of the system. Systems may also be created based on commercial, widely used program system GIS. Then into the created solution it is possible to implement modules that accept also data in other standards and secure their conversion into a used data standard. These are for instance data in Keyhole Markup Language format [KML, see OGC (2016)], open source CAD data, etc.

Data independence on program equipment is important on both sides also due to a possible change of the used program equipment for the given technology. This is supported by the fact that data provide much bigger inertia than program equipment.

Data independence on the program equipment then represents clear rules defining used standards that are respected by both producers as well as creators of application program equipment. In this respect, it makes sense to evaluate criterion k_{52} only in a way whether x-units of measure of database do or do not depend on the program equipment, thus:

- $k_{52} = 0$, if data that depend on program equipment of data producer are provided,
- $k_{52} = 100$, if data that depend on program equipment of data producer are not provided.

Data security represents a system of data security against unintentional or intentional damage, abuse, or loss. Within this criterion, it is necessary to evaluate the individual components of this system which in terms of work purpose relate only to utility characteristics of digital data. Components which provide data security in terms of production technologies are not evaluated here.

Data security from the user's point of view is done by:

- Security of user's rights to access databases,
- System of user's right security,
- Security of the whole system of data transfer from the data producer to the user including their physical security.

Each partial component may be evaluated by a level of security in a hundred-point scale where 100 means full protection and coefficient α then a deduction for not meeting the criterion; i-partial component of x-unit of measure of the database is then evaluated:

$$k_{53i} = 100 - \alpha_{53i}. \tag{4.39}$$

Assuming that all partial components of criterion k_{53} are of the same weight when securing the complete data protection, it is possible to express this for x-unit of measure of the database:

$$k_{53} = \frac{\sum_{i=1}^{n} k_{53i}}{n}, \tag{4.40}$$

where n is the number of all partial components of the criterion.

The total value of criterion k_5 may then be expressed by a function:

$$k_5 = \frac{\sum_{i=1}^{3} p_{5i} k_{5i}}{\sum_{i=1}^{3} p_{5i}}. \tag{4.41}$$

4.3 Utility Value of Digital Geographic Information

Digital geographic information is usually created and maintained in the individual partial components of the whole database, for instance in units of measure, saving units, map sheets, restricted areas of interest, etc. Due to this it is good to evaluate the individual utility value, for the partial components of the database, by a system described in the previous paragraph. As already mentioned, the whole database may be taken as a unit of measure, especially if it is created on purpose for a solution of chosen tasks of a smaller extent, or its system enables to work with the complete database at once.

4.3.1 Weights of Partial Criteria

In order to calculate utility as well as individual utility values of DGI, or their partial components, it is necessary to know weights of the partial criteria for the individual main criteria. The value of these weights may be defined based on a survey with data users. As an example, we may give weights of criteria obtained by a survey of DMU25 users (Talhofer, 2002).

The resulting weight values were re-calculated for the individual products with the use of weights of the main criteria by the method of gradual weight decomposition (Table 4.5).

4.3.2   Construction of Expression of the Individual Utility Value of DMU

When evaluating database utility, first it is good to define the ideal level of quality. This ideal level serves as a comparative standard for the expression of level of fulfillment of the individual criteria. Using the comparative standard, it is possible to evaluate the level of fulfillment of the individual criteria and subsequently also the total utility value.

The level of fulfillment of the individual criteria in general may be expressed as:

$$u_s = \frac{k_s}{\bar{k}_s},$$
(4.42)

where

– k_s is the value of fulfillment of s-partial criterion,
– $\bar{k}_s$ is the value of fulfillment of s-partial criterion or criterion of its sub-group of the comparative standard.

The level of fulfillment of the individual criteria must be calculated with regards to their structure, i.e. to their division to partial criteria of 1st, or 2nd and 3rd group.

4.3.2.1   Level of Fulfillment of Criterion "Content of Database"

The ideal value of fulfillment of criterion "*content of database*" k_1 is reached if the model of the real world is complete from the point of view of all users and all required levels of detail of data are kept. Thus, if the level of discrepancy with user requirements $\alpha_{11} = 0$,

– $n_{12i} = n_d$, $i = 1, 2$.

Ideal values of the individual criteria are:

– $\bar{k}_{11} = 100$,
– $\bar{k}_{12i} = 100$, $i = 1, 2$.

Then levels of fulfillment of the individual partial criteria for database u_i may be expressed:

$$u_{11} = 1 - \frac{\alpha_{11}}{100}, \ u_{12i} = \frac{n_{12i}}{n_d}.$$

The level of fulfillment of partial criterion u_1 *content of database* is then calculated from relation:

$$u_1 = \frac{p_{11}u_{11} + \sum_{i=1}^{2} p_{12i}u_{12i}}{p_{11} + \sum_{i=1}^{2} p_{12i}}.$$
(4.43)

4.3.2.2 Level of Fulfillment of Criteria "Technical Quality of Database Content"

Criterion k_2 *technical quality of database content* will have an ideal value in case that all data is flawless, accurate and consistent. The individual parameters of partial criteria must get the following values:

- $\alpha_{21i} = 0$, $i = 1, 2$,
- $100\frac{n_{22i}}{n} \geq 100 - h_s$, $i = 1, 2$,
- $100\frac{n_{23}}{n} \geq 100 - h_s$,
- $n_{24i} = n$, $i = 1, \ldots, 4$,
- $n_{25i} = n$, $i = 1, 2$.

Under these assumptions, the ideal values of the individual partial criteria will always equal 100. Then it is possible to calculate the level of fulfillment of the individual partial criteria for x-unit of measure for the individual components of the criterion according to the following relations:

- Transparency of data sources and methods for secondary data derivation:

$$u_{21i}^x = 1 - \frac{\alpha_{21i}^x}{100}, \quad i = 1, 2; \text{ and from that}$$

$$u_{21}^x = 1 - \frac{\sum_{i=1}^{2} p_{21i} \frac{\alpha_{21i}^x}{100}}{\sum_{i=1}^{2} p_{21i}}. \tag{4.44}$$

- Positional accuracy of data:

$$u_{22i}^x = \frac{n_{22i}^x}{n^x} + \frac{h_s}{100}, \quad i = 1, 2; \text{ then}$$

$$u_{22}^x = \frac{\sum_{i=1}^{2} p_{22i}\left(\frac{n_{22i}^x}{n^x} + \frac{h_s}{100}\right)}{\sum_{i=1}^{2} p_{22i}}. \tag{4.45}$$

- Attribute accuracy:

$$u_{23}^x = \frac{n_{23}^x}{n^x} + \frac{h_s}{100}. \tag{4.46}$$

- Logical consistency of the database:

$$u_{24i}^x = \frac{n_{24i}^x}{n^x}, \text{ from which we get}$$

$$u_{24}^x = \frac{\sum_{i=1}^3 p_{24i} \frac{n_{24i}^x}{n^x}}{\sum_{i=1}^3 p_{24i}}. \tag{4.47}$$

– Data completeness:

$$u_{25}^x = \frac{\sum_{i=1}^2 p_{25i} \frac{n_{25i}^x}{n^x}}{\sum_{i=1}^2 p_{25i}}. \tag{4.48}$$

The total value of level of fulfillment of criterion *technical quality of database content* u_2 is then expressed from the individual partial criteria:

$$u_2^x = \frac{\sum_{i=1}^5 p_{2i} u_{2i}^x}{\sum_{i=1}^5 p_{2i}}. \tag{4.49}$$

Once the values of the individual partial criteria have been substituted, relation (4.49) may be expressed:

$$u_2^x = \frac{p_{21} - p_{21} \frac{\sum_{i=1}^2 p_{21i} \frac{\alpha_{21i}^x}{100}}{\sum_{i=1}^2 p_{21i}} + p_{22} \left[\frac{\sum_{i=1}^2 p_{21i} \left(\frac{n_{22i}^x}{n^x} + \frac{h_s}{100} \right)}{\sum_{i=1}^2 p_{22i}} \right]}{\sum_{i=1}^5 p_{22i}}$$
$$+ \frac{p_{23} \left(\frac{n_{23}^x}{n^x} + \frac{h_s}{100} \right) + p_{24} \left(\frac{\sum_{i=1}^3 p_{24i} \frac{n_{24i}^x}{n^x}}{\sum_{i=1}^3 p_{24i}} \right) + p_{25} \left(\frac{\sum_{i=1}^2 p_{25i} \frac{n_{25i}^x}{n^x}}{\sum_{i=1}^2 p_{25i}} \right)}{\sum_{i=1}^5 p_{22i}} . \tag{4.50}$$

4.3.2.3   Level of Fulfillment of Criterion "Database Timeliness"

Criterion *database timeliness* k_3 will have the ideal value in case the database is used in time $\bar{T}_{ed}$, for which the given conformity of the modelled reality and the content of database applies. The ideal value $\bar{k}_3$ may be expressed as:

$$\bar{k}_3 = \bar{f}_i(\bar{T}_{ed}), \quad i \in 1, 2, \ldots, 4.$$

From the formulas of all functions for criteria expression it is obvious that the ideal value for all functions equals 1. It means that the value of level of fulfillment of criteria u_3 *database timeliness* of x-unit of measure may be expressed by the same functions as in Sect. 4.2.2.2, i.e.:

$$u_3^x = \frac{C(T^x - \bar{T}_{ed}) - (T^x - T_{lim})}{T_{lim} - \bar{T}_{ed}}, \tag{4.51}$$

or

$$u_3^x = e^{-b(T^x - \bar{T}_{ed})},\tag{4.52}$$

or

$$u_3^x = \left[1 - \frac{1}{T_{lim}}\right]^{T^x - \bar{T}_{ed}},\tag{4.53}$$

or

$$u_3^x = (1 - C)\left[1 - \frac{1}{T_{lim}}\right]^{T^x - \bar{T}_{ed}} + C.\tag{4.54}$$

4.3.2.4 Level of Fulfillment of Criterion "Area Importance"

When evaluating the level of criterion k_4 *area importance* it is not possible to say exactly what the ideal value of this criterion is. In sense of criteria analysis in Sect. 4.2.2.2 it is obvious that some of the partial criteria are invariant to user's needs—e.g. geographic position, existence of settlement units, etc. Yet also for this criterion for a general function of the designed system it is necessary to evaluate the level of its fulfillment. The ideal value will then be, if the area is maximally important for the user, i.e. if value $\bar{k}_4 = 1$. The level of its fulfillment may be calculated with the help of equation:

$$u_4^x = p_1 k_{41}^x \sum_{i=2}^{n} p_i k_{4i}^x.\tag{4.55}$$

4.3.2.5 Level of Fulfillment of Criterion "User Friendliness"

The ideal value of criterion k_5 *user friendliness* happens when all data comply with the given standard, they are generally usable in various program systems without the necessity to modify them beforehand and they are secured against abuse and breach of copyright. In such case the values of partial parameters will be:

- $\bar{k}_{5i} = 100$, $i = 1, 2$,
- $\alpha_{53i} = 0$, $i = 1 \ldots 3$.

For x-unit of measure it will apply that:

$$u_5^x = \frac{p_{51}\frac{k_{51}^x}{100} + p_{52}\frac{k_{52}^x}{100} + \left(1 - \frac{\sum_{i=1}^{3} p_{53i}\frac{\alpha_{53i}^x}{100}}{\sum_{i=1}^{3} p_{53i}}\right)}{\sum_{i=1}^{3} p_{5i}}.\tag{4.56}$$

4.3.3   Expression of the Complete Individual Utility Value

The complete *individual utility value U* or *individual functionality F* of *x*-unit of measure is given by an aggregation function of the same kind which may be expressed as:

$$U^x = F^x = p_3 u_3^x p_4 u_4^x \left(p_1 u_1^x + p_2 u_2^x + p_5 u_5^x \right). \tag{4.57}$$

The weights of the individual criteria are identical to weights for the calculation of utility value of the database and it is possible to use Table 4.12. The formula of the resulting function is chosen with regards to the usability of DGI database. The user may not be satisfied with the definition of content, technical quality or level of user friendliness, but still they can be able to work with the given database, even though not fully. However, if the database is absolutely obsolete or is provided from an area beyond their AOI, this database becomes totally worthless.

The Eq. (4.57) for calculation of the complete individual utility value U is a function of total of 29 variables which characterize the level of fulfillment of the individual criteria. Under the assumption that the individual variables depend on each other, it is possible to model the mentioned function as well as changes of the individual utility value and thus evaluate which database characteristics are crucial for the given usage and which influence the individual utility value less. This way it is then possible to create variants of the product and together with expenses for the provision of the functionality to optimize the creation or update of the given database or its partial component.

4.3.4   Influence of a Change of Functionally Conditioned Characteristics on the Utility Value of DGI

Changes of characteristics of DGI manifest in a change of the utility value of this database. It is possible to define this change based on an analysis of changes of values of the evaluated criteria. To secure their influence on the complete utility value it is possible to use derivation function F according to the individual variables expressing the level of fulfillment of the given criteria. With regards to the function formula (4.57), in general it is possible to express the influence of a change of level of fulfillment of *i*-main criterion as:

$$F' = \frac{dU}{du_i}. \tag{4.58}$$

However, the levels of fulfillment of main criteria are a function of more variables. It is possible to approach the expression of value du_i in two ways depending on the required structure of information. If it is necessary to evaluate *the influence of the*

Table 4.12 Weights of partial criteria as average values obtained from the survey

Criterion weight	p_i	p_{ij}	p_{ijk}	p
p_1	0.233			0.233
p_{11}		0.534		0.124
p_{12}		0.466		0.109
p_{121}			0.517	0.056
p_{122}			0.483	0.053
p_2	0.201			0.201
p_{21}		0.218		0.044
p_{211}			0.547	0.024
p_{212}			0.453	0.020
p_{22}		0.219		0.044
p_{221}			0.570	0.025
p_{222}			0.430	0.019
p_{23}		0.175		0.035
p_{24}		0.195		0.039
p_{241}			0.432	0.017
p_{242}			0.344	0.014
p_{243}			0.223	0.009
p_{25}		0.193		0.039
p_{251}			0.519	0.020
p_{252}			0.481	0.019
p_3	0.219			0.219
p_4	0.198			0.198
p_{41}		0.143		0.028
p_{42}		0.096		0.019
p_{43}		0.101		0.020
p_{44}		0.113		0.022
p_{45}		0.091		0.018
p_{46}		0.081		0.016
p_{47}		0.079		0.016
p_{48}		0.088		0.017
p_{49}		0.114		0.023
p_{4_10}		0.094		0.019
p_5	0.148			0.148
p_{51}		0.369		0.055
p_{52}		0.306		0.045
p_{53}		0.325		0.048
p_{531}			0.314	0.015
p_{532}			0.331	0.016
p_{533}			0.356	0.017

individual variables on the complete individual utility value when assuming that values of all other variables are constant, then it is necessary to express derivations of function U as:

$$F' = \frac{dU}{du_i} \frac{du_i}{dx},$$

(4.59)

where x is one of the mentioned 29 variables.

However, in practice we may encounter a situation when there are changes of more factors happening at the same time. For instance, the technical quality of database content changes in all parameters—used methods of deriving secondary data improves, positional and attribute accuracy and data complexity increases, and moreover, data is stored in a geodatabase accessible to authorized users which contains all quality secured topological, thematic and time relations. In this case, it is good to express value du_i as a total differentiation of all variables. The following paragraphs give results of both approaches.

4.3.4.1 Influence of Changes of Database Content

In the first case, influences of the individual partial criteria of criterion *content of database* on the complete individual utility value are evaluated. For the individual variables, it is possible—using formulas (4.43), (4.57), and (4.59)—to express for x-unit of measure of database:

$$\frac{dU^x}{d\alpha_{11}} = \frac{dU^x}{du_1} \frac{du_1}{d\alpha_{11}} = p_1 p_3 u_3^x p_4 u_4^x \left[-\frac{p_{11}}{100(p_{11} + \sum_{i=1}^2 p_{21i})} \right],$$

(4.60)

$$\frac{dU^x}{dn_{121}} = \frac{dU^x}{du_1} \frac{du_1}{dn_{121}} = \frac{p_1 p_3 u_3^x p_4 u_4^x p_{121}}{n_d(p_{11} + \sum_{i=1}^2 p_{12i})},$$

(4.61)

$$\frac{dU^x}{dn_{122}} = \frac{dU^x}{du_1} \frac{du_1}{dn_{122}} = \frac{p_1 p_3 u_3^x p_4 u_4^x p_{122}}{n_d(p_{11} + \sum_{i=1}^2 p_{12i})}.$$

(4.62)

In case the values of changes of the individual variables in formulas (4.60), (4.61) and (4.62) get final values, then the mentioned formulas may be expressed:

$$\Delta U^x = p_1 p_3 u_3^x p_4 u_4^x \left[-\frac{p_{11}}{100(p_{11} + \sum_{i=1}^2 p_{12i})} \right] \Delta \alpha_{11},$$

(4.63)

$$\Delta U^x = \frac{p_1 p_3 u_3^x p_4 u_4^x p_{122}}{n_d(p_{11} + \sum_{i=1}^2 p_{12i})} \Delta n_{121},$$

(4.64)

$$\Delta U^x = \frac{p_1 p_3 u_3^x p_4 u_4^x p_{122}}{n_d(p_{11} + \sum_{i=1}^2 p_{12i})} \Delta n_{122}.$$

(4.65)

Increases Δn_{121}, Δn_{122} are expressed in absolute values.

In the other case, when we consider the influence of more individual criteria at the same time, firstly the overall influence of the individual variables of the main criterion u_1 on the complete individual utility value will be evaluated. The complete change of level of fulfillment of criterion u_1 may be expressed as a total derivative of function (4.43), i.e.:

$$du_1 = \frac{\partial u_1}{\partial \alpha_{11}} d\alpha_{11} + \frac{\partial u_1}{\partial n_{121}} dn_{121} + \frac{\partial u_1}{\partial n_{122}} dn_{122}.$$

After adjustment, we get the relation:

$$du_1 = \frac{p_{121}dn_{121} + p_{122}dn_{122}}{n_d(p_{11} + \sum_{i=1}^{2} p_{12i})} - \frac{p_{11}d\alpha_{11}}{100(p_{11} + \sum_{i=1}^{2} p_{12i})}. \tag{4.66}$$

The complete influence on the individual utility value of x-unit of measure of database is then calculated with the help of formulas (4.57), (4.58) and (4.66) as:

$$dU^x = \left[\frac{p_{121}dn_{121} + p_{122}dn_{122}}{n_d(p_{11} + \sum_{i=1}^{2} p_{12i})} - \frac{p_{11}d\alpha_{11}}{100(p_{11} + \sum_{i=1}^{2} p_{12i})} \right] p_1 p_3 u_3^x p_4 u_4^x. \tag{4.67}$$

Alternatively, for final values of increases of the individual variables:

$$\Delta U^x = \left[\frac{p_{121}\Delta n_{121} + p_{122}\Delta n_{122}}{n_d(p_{11} + \sum_{i=1}^{2} p_{12i})} - \frac{p_{11}\Delta\alpha_{11}}{100(p_{11} + \sum_{i=1}^{2} p_{12i})} \right] p_1 p_3 u_3^x p_4 u_4^x. \tag{4.68}$$

4.3.4.2 Influence of a Change of Technical Quality of Database Content

As well as in Sect. 4.3.4.1, using formula (4.50) for the individual variables of main criterion u_2 it is possible to express their influence on the complete individual utility value of x-unit of measure of database with the following relations:

$$\frac{dU^x}{d\alpha_{211}^x} = \frac{dU^x}{du_2^x}\frac{du_2^x}{d\alpha_{211}^x} = -p_1 p_3 u_3^x p_4 u_4^x \frac{p_{21}p_{211}}{100 \sum_{i=1}^{2} p_{21i} \sum_{i=1}^{5} p_{2i}}, \tag{4.69}$$

$$\frac{dU^x}{d\alpha_{212}^x} = \frac{dU^x}{du_2^x}\frac{du_2^x}{d\alpha_{212}^x} = -p_2 p_3 u_3^x p_4 u_4^x \frac{p_{21}p_{212}}{100 \sum_{i=1}^{2} p_{21i} \sum_{i=1}^{5} p_{2i}}, \tag{4.70}$$

$$\frac{dU^x}{dn_{211}^x} = \frac{dU^x}{du_2^x}\frac{du_2^x}{dn_{211}^x} = p_2 p_3 u_3^x p_4 u_4^x \frac{p_{22}p_{221}}{100 \sum_{i=1}^{2} p_{22i} \sum_{i=1}^{5} p_{2i}}, \tag{4.71}$$

$$\frac{dU^x}{dn_{222}^x} = \frac{dU^x}{du_2^x}\frac{du_2^x}{dn_{222}^x} = p_2 p_3 u_3^x p_4 u_4^x \frac{p_{22}p_{222}}{n^x \sum_{i=1}^{2} p_{22i} \sum_{i=1}^{5} p_{2i}}, \tag{4.72}$$

$$\frac{dU^x}{dn_{23}^x} = \frac{dU^x}{du_2^x}\frac{du_2^x}{dn_{23}^x} = p_2 p_3 u_3^x p_4 u_4^x \frac{p_{23}}{n^x \sum_{i=1}^{5} p_{2i}}, \tag{4.73}$$

$$\frac{dU^x}{dn^x_{241}} = \frac{dU^x}{du^x_2}\frac{du^x_2}{dn^x_{241}} = p_2 p_3 u^x_3 p_4 u^x_4 \frac{p_{24} p_{241}}{n^x \sum^2_{i=1} p_{24i} \sum^5_{i=1} p_{2i}},\tag{4.74}$$

$$\frac{dU^x}{dn^x_{242}} = \frac{dU^x}{du^x_2}\frac{du^x_2}{dn^x_{242}} = p_2 p_3 u^x_3 p_4 u^x_4 \frac{p_{24} p_{242}}{n^x \sum^3_{i=1} p_{24i} \sum^5_{i=1} p_{2i}},\tag{4.75}$$

$$\frac{dU^x}{dn^x_{243}} = \frac{dU^x}{du^x_2}\frac{du^x_2}{dn^x_{243}} = p_2 p_3 u^x_3 p_4 u^x_4 \frac{p_{24} p_{243}}{n^x \sum^3_{i=1} p_{24i} \sum^5_{i=1} p_{2i}},\tag{4.76}$$

$$\frac{dU^x}{dn_{251}{}^x} = \frac{dU^x}{du^x_2}\frac{du_2{}^x}{dn^x_{251}} = p_2 p_3 u^x_3 p_4 u^x_4 \frac{p_{25} p_{251}}{n^x \sum^2_{i=1} p_{25i} \sum^5_{i=1} p_{2i}},\tag{4.77}$$

$$\frac{dU^x}{dn_{252}{}^x} = \frac{dU^x}{du^x_2}\frac{du_2{}^x}{dn^x_{252}} = p_2 p_3 u^x_3 p_4 u^x_4 \frac{p_{25} p_{252}}{n^x \sum^2_{i=1} p_{25i} \sum^5_{i=1} p_{2i}}.\tag{4.78}$$

The overall influence of all variables of criterion *technical quality of database content* u_2 may be expressed again by the total differentiation:

$$\begin{aligned}
du_2 = & \frac{\partial u_2}{\partial \alpha_{211}}d\alpha_{211} + \frac{\partial u_2}{\partial \alpha_{212}}d\alpha_{212} + \frac{\partial u_2}{\partial \alpha_{221}}d\alpha_{221} + \frac{\partial u_2}{\partial \alpha_{222}}d\alpha_{222}\\
& + \frac{\partial u_2}{\partial n_{23}}dn_{23} + \frac{\partial u_2}{\partial n_{241}}dn_{241} + \frac{\partial u_2}{\partial n_{242}}dn_{242}\\
& + \frac{\partial u_2}{\partial n_{243}}dn_{243} + \frac{\partial u_2}{\partial n_{251}}dn_{251} + \frac{\partial u_2}{\partial n_{252}}dn_{252}
\end{aligned}\tag{4.79}$$

After substitution of results of the individual partial derivations, adjustment of differentiations of the individual variables to the final values and substitution of the absolute numbers of their percentage expression, Eq. (4.79) may be written in this form:

$$\begin{aligned}
\Delta u_2 = & -\frac{p_{21}\sum^2_{i=1}p_{21i}\Delta\alpha_{21i}}{100\sum^2_{i=1}p_{21i}\sum^5_{i=1}p_{2i}} + \frac{p_{22}\sum^2_{i=1}p_{22i}\Delta n'_{22i}}{100\sum^2_{i=1}p_{22i}\sum^5_{i=1}p_{2i}} + \frac{p_{23}\Delta n'_{23}}{100\sum^2_{i=1}p_{2i}}\\
& + \frac{p_{24}\sum^3_{i=1}p_{24i}\Delta n'_{24i}}{100\sum^2_{i=1}p_{24i}\sum^5_{i=1}p_{2i}} + \frac{p_{25}\sum^2_{i=1}p_{25i}\Delta n'_{25i}}{100\sum^2_{i=1}p_{25i}\sum^5_{i=1}p_{2i}}
\end{aligned}\tag{4.80}$$

The resulting change of the individual utility value of x-unit of measure of database may then be calculated using Eqs. (4.57) and (4.80).

$$\begin{aligned}
\Delta U^x = & p_2 p_3 u^x_3 p_4 u^x_4\left(-\frac{p_{21}\sum^2_{i=1}p_{21i}\Delta\alpha^x_{21i}}{100\sum^2_{i=1}p_{21i}\sum^5_{i=1}p_{2i}} + \frac{p_{22}\sum^2_{i=1}p_{22i}\Delta n'^x_{22i}}{100\sum^2_{i=1}p_{22i}\sum^5_{i=1}p_{2i}} + \frac{p_{23}\Delta n'^x_{23}}{100\sum^2_{i=1}p_{2i}}\right.\\
& \left.+ \frac{p_{24}\sum^3_{i=1}p_{24i}\Delta n'^x_{24i}}{100\sum^2_{i=1}p_{24i}\sum^5_{i=1}p_{2i}} + \frac{p_{25}\sum^2_{i=1}p_{25i}\Delta n'^x_{25i}}{100\sum^2_{i=1}p_{25i}\sum^5_{i=1}p_{2i}}\right)
\end{aligned}\tag{4.81}$$

Criterion k_2 *technical quality of database content* is rather highly structured into another two levels. In practice there may be a situation when it is necessary to evaluate the influence of only variables characterizing a partial criterion of 1st group. For instance, it will be necessary to evaluate the influence of source transparency and methods of derivation of secondary data on the overall individual utility value, without evaluating the influence of other variables. The solution of such requirement lies in the calculation of total differentiations of functions characterizing levels of fulfillment of the individual partial criteria of 1st group. Thus for 1st partial criterion we may write:

$$\frac{dU}{du_{21}} = \frac{dU}{du_2}\frac{du_2}{du_{21}},$$

where

$$du_{21} = \frac{\partial u_{21}}{\partial \alpha_{211}}d\alpha_{211} + \frac{\partial u_{21}}{\partial \alpha_{212}}d\alpha_{212}.$$

Similarly, it is possible to express total differentiations of all other functions (4.45), (4.46), (4.47) and (4.48). Again assuming that differentiations of the individual variables will get final values and that variables expressing numbers will be transformed to their percentage values, it is possible to overall evaluate a change of level of fulfillment of criterion of 1st group of x-unit of measure of database according to relations:

$$\Delta U^x = -p_2 p_3 u_3^x p_4 u_4^x \left(\frac{p_{21} \sum_{i=1}^{2} p_{21i} \Delta \alpha_{21i}^x}{100 \sum_{i=1}^{2} p_{21i} \sum_{i=1}^{2} p_{2i}} \right), \tag{4.82}$$

$$\Delta U^x = p_2 p_3 u_3^x p_4 u_4^x \left(\frac{p_{22} \sum_{i=1}^{2} p_{22i} \Delta n_{22i}^{'x}}{100 \sum_{i=1}^{2} p_{22i} \sum_{i=1}^{2} p_{2i}} \right), \tag{4.83}$$

$$\Delta U^x = p_2 p_3 u_3^x p_4 u_4^x \left(\frac{p_{23} \Delta n_{23}^{'x}}{100 \sum_{i=1}^{5} p_{2i}} \right), \tag{4.84}$$

$$\Delta U^x = p_2 p_3 u_3^x p_4 u_4^x \left(\frac{p_{24} \sum_{i=1}^{3} p_{24i} \Delta n_{24i}^{'x}}{100 \sum_{i=1}^{3} p_{24i} \sum_{i=1}^{5} p_{2i}} \right), \tag{4.85}$$

$$\Delta U^x = p_2 p_3 u_3^x p_4 u_4^x \left(\frac{p_{25} \sum_{i=1}^{2} p_{25i} \Delta n_{25i}^{'x}}{100 \sum_{i=1}^{2} p_{25i} \sum_{i=1}^{5} p_{2i}} \right). \tag{4.86}$$

4.3.4.3 Influence of a Change of Database Timeliness

A change of the complete individual utility value given by the influence of a change of level of fulfillment of criterion *database timeliness-u_3* depends only on one variables—time T. That is the reason, why it is possible to express this change by relation:

$$\frac{dU}{dT} = \frac{dU}{du_3}\frac{du_3}{dT},$$

or

$$\frac{dU}{dT} = p_3 p_4 u_4 (p_1 u_1 + p_2 u_2 + p_5 u_5)\frac{du_3}{dT}. \tag{4.87}$$

As level u_3 of fulfillment of criterion k_3 is expressed in Sect. 4.3.2.3 by four equations [from (4.51) to (4.54)], its change may also be expressed with the help of 4 expressions:

$$\frac{du_3}{dT} = \frac{C-1}{T_{lim}-T_{ed}}, \tag{4.88}$$

or

$$\frac{du_3}{dT} = -be^{-b(T-T_{ed})}, \tag{4.89}$$

or

$$\frac{du_3}{dT} = \left(1-\frac{1}{T_{lim}}\right)^{T-T_{ed}}\ln\left(1-\frac{1}{T_{lim}}\right), \tag{4.90}$$

or

$$\frac{du_3}{dT} = (1-C)\left(1-\frac{1}{T_{lim}}\right)^{T-T_{ed}}\ln\left(1-\frac{1}{T_{lim}}\right). \tag{4.91}$$

Assuming that differentiation value of time change replaces the final value ΔT, it is possible for x-unit of measure of database to evaluate change of its individual utility value depending on time by equation:

The total change of utility value of x-unit of measure is gained by substituting one of the expressions from (4.88) to (4.91) into expression (4.87) and again

$$\Delta U^x = p_3 p_4 u_4^x (p_1 u_1 + p_2 u_2^x + p_5 u_5^x)\frac{du_3}{dT}\Delta T^x. \tag{4.92}$$

The choice of an Eq. (4.88) to (4.91) again depends on the stability of the area with regard to time and on the type of changes happening in the area.

4.3.4.4   Influence of a Change of Area Importance

Area importance is given by many factors (see Sect. 4.2.2.2) whose influence on the total individual value may be discovered but which a geographer as a database producer usually cannot affect. If these factors change in course of time or the position of the area of interest changes, this change always manifests in the complete level of fulfillment of criterion-u_4. That is the reason why it is good to evaluate only the influence of the complete change of area importance on the complete individual value which may be expressed by relation:

$$\mathrm{d}F = \frac{\mathrm{d}U}{\mathrm{d}u_4}.$$

If the differentiation of a change of level of fulfillment of criterion u_4 is substituted by the final value, for x-unit of measure of database, it will apply:

$$\Delta U^x = p_3 u_3^x p_4 \left(p_1 u_1 + p_2 u_2^x + p_5 u_5^x \right) \Delta u_4^x. \tag{4.93}$$

4.3.4.5 Influence of a Change of User Friendliness

The influence of a change of level of fulfillment of criterion *user friendliness* on the total individual utility value of x-unit of measure of database may be expressed again either as an influence of the individual variables of criterion k_5 with the help of its differentiations, or their overall influence with the help of a total differentiation of function (4.56). In the former case, the influence of the individual variables is expressed using the following relations:

$$\frac{\mathrm{d}U^x}{\mathrm{d}k_{51}^x} = \frac{\mathrm{d}U^x}{\mathrm{d}u_5^x}\frac{\mathrm{d}u_5^x}{\mathrm{d}k_{51}^x} = p_5 p_3 u_3^x p_4 u_4^x \frac{p_{51}}{100 \sum_{i=1}^{3} p_{5i}}, \tag{4.94}$$

$$\frac{\mathrm{d}U^x}{\mathrm{d}k_{52}^x} = \frac{\mathrm{d}U^x}{\mathrm{d}u_5^x}\frac{\mathrm{d}u_5^x}{\mathrm{d}k_{52}^x} = p_5 p_3 u_3^x p_4 u_4^x \frac{p_{52}}{100 \sum_{i=1}^{3} p_{5i}}, \tag{4.95}$$

$$\frac{\mathrm{d}U^x}{\mathrm{d}\alpha_{531}^x} = \frac{\mathrm{d}U^x}{\mathrm{d}u_5^x}\frac{\mathrm{d}u_5^x}{\mathrm{d}\alpha_{531}^x} = -p_5 p_3 u_3^x p_4 u_4^x \frac{p_{53}p_{531}}{100 \sum_{i=1}^{3} p_{53i} \sum_{i=1}^{3} p_{5i}}, \tag{4.96}$$

$$\frac{\mathrm{d}U^x}{\mathrm{d}\alpha_{532}^x} = \frac{\mathrm{d}U^x}{\mathrm{d}u_5^x}\frac{\mathrm{d}u_5^x}{\mathrm{d}\alpha_{532}^x} = -p_5 p_3 u_3^x p_4 u_4^x \frac{p_{53}p_{532}}{100 \sum_{i=1}^{3} p_{53i} \sum_{i=1}^{3} p_{5i}}, \tag{4.97}$$

$$\frac{\mathrm{d}U^x}{\mathrm{d}\alpha_{533}^x} = \frac{\mathrm{d}U^x}{\mathrm{d}u_5^x}\frac{\mathrm{d}u_5^x}{\mathrm{d}\alpha_{533}^x} = -p_5 p_3 u_3^x p_4 u_4^x \frac{p_{53}p_{533}}{100 \sum_{i=1}^{3} p_{53i} \sum_{i=1}^{3} p_{5i}}. \tag{4.98}$$

Just like in all previous cases it is good in the preceding calculations to substitute differentiation values of the individual variables by their final increases. Then the equations look like this:

$$\Delta U^x = p_5 p_3 u_3^x p_4 u_4^x \frac{p_{51}}{100 \sum_{i=1}^{3} p_{5i}} \Delta k_{51}^x, \tag{4.99}$$

$$\Delta U^x = p_5 p_3 u_3^x p_4 u_4^x \frac{p_{52}}{100 \sum_{i=1}^{3} p_{5i}} \Delta k_{52}^x, \tag{4.100}$$

$$\Delta U^x = -p_5 p_3 u_3^x p_4 u_4^x \frac{p_{53} p_{531}}{100 \sum_{i=1}^{3} p_{53i} \sum_{i=1}^{3} p_{5i}} \Delta \alpha_{531}^x, \qquad (4.101)$$

$$\Delta U^x = -p_5 p_3 u_3^x p_4 u_4^x \frac{p_{53} p_{532}}{100 \sum_{i=1}^{3} p_{53i} \sum_{i=1}^{3} p_{5i}} \Delta \alpha_{532}^x, \qquad (4.102)$$

$$\Delta U^x = -p_5 p_3 u_3^x p_4 u_4^x \frac{p_{53} p_{533}}{100 \sum_{i=1}^{3} p_{53i} \sum_{i=1}^{3} p_{5i}} \Delta \alpha_{533}^x. \qquad (4.103)$$

The overall influence of all variables of criterion k_5 may be expressed as total differentiation of function (4.56):

$$\frac{dU}{du_5} = \frac{\partial u_5}{\partial u_{51}} dk_{51} + \frac{\partial u_5}{\partial u_{52}} dk_{52} + \frac{\partial u_5}{\partial \alpha_{531}} d\alpha_{531} + \frac{\partial u_5}{\partial \alpha_{532}} d\alpha_{532} + \frac{\partial u_5}{\partial \alpha_{533}} d\alpha_{533}. \quad (4.104)$$

If the differentiations of the individual variables are substituted by their final values, then for x-unit of measure it is possible to write:

$$\Delta U^x = p_5 p_3 u_3^x p_4 u_4^x \left(\frac{p_{51} \Delta k_{51}^x + p_{52} \Delta k_{52}^x}{100 \sum_{i=1}^{3} p_{5i}} - \frac{p_{53} \sum_{i=1}^{3} p_{53i} \Delta \alpha_{53i}^x}{100 \sum_{i=1}^{3} p_{5i} \sum_{i=1}^{3} p_{53i}} \right). \quad (4.105)$$

4.4 The Use of Value Analysis in Order to Improve Utility of Products of Geographic Service

The previous mathematical model may be used for modelling of utility and economic efficiency of current—as well as newly created—products of geographic service. The model can be used as a whole when controlling workers may examine behavior of the complete utility value or individual utility value of the given database of area digital model while changing settings of the individual parameters of the model; it may also be used with its individual parts when it is possible to observe the tendency of influence of the individual evaluated parameters on the complete utility value if these parameters change in beforehand given limits. At the same time, it is possible to observe the economic efficiency with the help of RCE indicator. The following example is an analysis of behavior of one unit of measure of DMU25 database in the range of 10 × 10 km.

4.4.1 Default Conditions of Analysis

Default parameters of DMU25 were adopted from catalogue of objects (MoD-GeoS, 2013). For setting parameters of evaluation functions of criterion *database timeliness*, the following model values relating to the territory of the Czech Republic were used:

Table 4.13 Resulting weights for DMU25

Weight of criterion	Database DMU25	Criterion weight	Database DMU25
p_{11}	0.124	p_3	0.747
p_{121}	0.056	p_{41}	0.408
p_{122}	0.053	p_{42}	0.274
p_1	0.794	p_{43}	0.289
p_{211}	0.024	p_{44}	0.324
p_{212}	0.020	p_{45}	0.259
p_{21}	0.044	p_{46}	0.233
p_{221}	0.025	p_{47}	0.227
p_{222}	0.019	p_{48}	0.251
p_{22}	0.044	p_{49}	0.327
p_{23}	0.035	p_{4_10}	0.267
p_{23}	0.035	p_4	0.674
p_{241}	0.017	p_{51}	0.055
p_{242}	0.014	p_{52}	0.045
p_{243}	0.009	p_{531}	0.015
p_{24}	0.039	p_{532}	0.016
p_{251}	0.020	p_{533}	0.017
p_{252}	0.019	p_{53}	0.048
p_{25}	0.039	p_5	0.505
p_2	0.686		

- T_{lim} = 10 years,
- T_{ed} = 1 year,
- $C = 0.5$,
- $d = 1.5$.

Weights for the individual criteria were adopted from Table 4.12. Weights that related to multiple parameters for calculations were recalculated according to appropriate equations. The resulting weights used are given in the Table 4.13.

Furthermore, the influence of the individual variables on functions expressing the level of fulfillment of the individual main criteria was investigated. This step was based on ideal values of all partial criteria and only one variable in the extent of values that this criterion may reach in real conditions was changed. The aim was to observe the behavior of the main criterion. Only resulting graphs presenting this part of modelling are shown.

In case of criterion *content of database*, interval 0–15%, resp. 85–100%, was chosen as a model range of values of the individual variables. It is obvious from the graph (Fig. 4.10) that the completeness of the real world model has the biggest

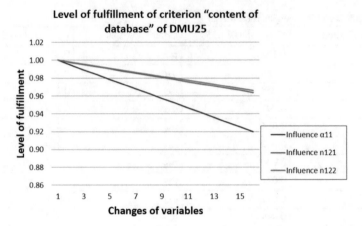

Fig. 4.10 Influence of individual variables on the level of fulfillment of criterion *content of database* of DMU25

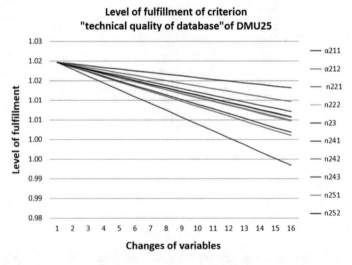

Fig. 4.11 Influence of individual variables on the level of fulfillment of criterion *technical quality of database content* of DMU25, all variables are changed in the same range

influence on the level of fulfillment of this criterion, the influence of keeping the required geometric as well as thematic level of detail is basically identical.

For criterion *technical quality of database content* for the individual variables, model values in the range of 0–15%, resp. 85–100%, were chosen once again. The graph (Fig. 4.11) presents the influence of these individual variables.

It is obvious from the graph that in this case the biggest influence comes from criterion *attributive accuracy*, then *positional accuracy*, and *transparency of source material on primary data collection*. The influence of a change of time passed from

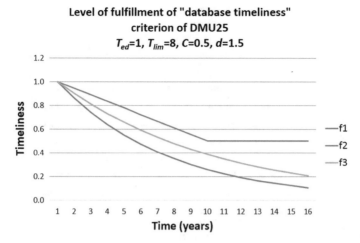

Fig. 4.12 Influence of form of the chosen function for expression of time change on the level of fulfillment of criterion *database timeliness*

the provision of database DMU25 to users on the behavior of functions expressing the course of obsolescence and thus also the shape of functions for calculation of level of fulfillment of criterion *database timeliness* is shown in the graph (Fig. 4.12). Only functions $f_1(T)$, $f_2(T)$ and $f_3(T)$ from Sect. 4.2.2.2 were used. Function $f_4(T)$ was not used as it is used first of all for an expression of very stable objects of databases which is not true in this case. The range of entry data (of previous years from value T_{lim}) was chosen in the interval of 1–16. The graphs clearly show that before making the decision which function to choose for the evaluation of the individual utility value it is necessary for the given area to make an analysis of frequency of changes of the modelled objects and at the same time to consider how the complete timeliness of database will change in this space.

The influence of a change of level of fulfillment of criterion *area importance* is given in a graph (Fig. 4.13). Modelled values of particular variables are in the interval <0.5, 1.0>. From the mentioned graph it is evident the value of partial criterion *geographic position* has the most influence. In the case the data are not provided from required area, the criterion value equals zero (variant 11). Other variables have in fact the same influence. Their detailed influence is presented in the picture (Fig. 4.14), where the influence of criterion *geographic position* was not considered.

The graph (Fig. 4.15) shows the result of modelling the change of level of fulfillment of criterion *user friendliness* due to changes of the individual variables of this criterion. In compliance with Sect. 4.2.2.2, for values of variables k_{51} and k_{52} only two values 100 and 0 were chosen, values of the other variables range from the interval of 0–15%. The graph clearly shows that having introduced standardization, the level of fulfillment of whole criterion k_5 significantly increased, as well as when introduced data independence on the concrete program environment. The other variables do not influence the level of fulfillment of this criterion much.

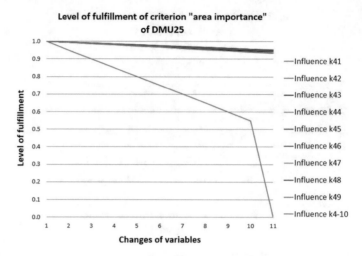

Fig. 4.13 Influence of the individual variables on the level of fulfillment of criterion *area importance*

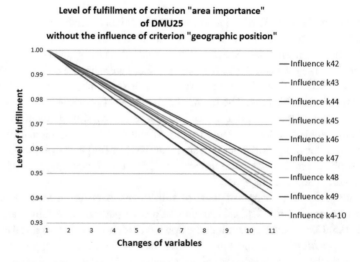

Fig. 4.14 Influence of the individual variables on the level of fulfillment of criterion *area importance* without influence of geographic position

Just like in the previous cases, it is possible to calculate influences of changes of all variables on the complete change of the individual utility value of database DMU25 calculated according to formula (4.57). The graph in the picture (Fig. 4.16) shows the tendency of change of this complete individual utility value with the exclusion of influence of a change of level of fulfillment of criteria *database timeliness, area importance* and *standardization, data independence and protection*. These criteria were considered to be ideally fulfilled with regards to the presumption that maximally

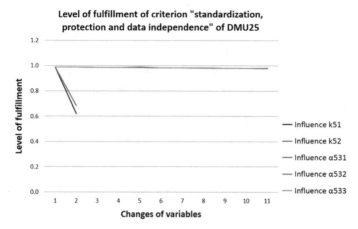

Fig. 4.15 Influence the individual variables on the level of fulfillment of criterion *standardization, data independence and protection*

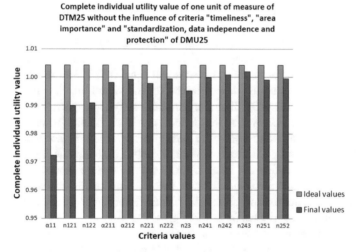

Fig. 4.16 Tendency of change of the complete individual utility value due to changes of variable criteria k_1 and k_2

important locality is evaluated, whose model that is just released and it is fully standardized does not depend on program equipment and it is maximally protected. The graph shows a situation that occurs with a gradual change of all other variables from ideal values to the worst. All variables in this case are alternatively 0 or 15% (resp. 100 or 85%).

It is clear from the graph that the level of criterion *content of database* has a significant influence on the complete individual utility value. If users' requirements on the content of database and on its level of detail are not respected, the utility value decreases much more than when its technical quality is increased.

4.4.2   Analysis of a Chosen Database Unit of Measure

The previous analysis only modelled influences of the individual variables on the complete individual utility value of database unit of measure (its functionality) on condition that other variables will have an "Ideal" value. However, this case will never happen in practice, at least because the geographic position of the area of interest will always change. This part of the analysis will then examine influences of a change of functionality for concrete (not ideal) values of individual variables.

4.4.3   The Degree of Fulfillment of Criteria and the Complete Individual Functionality

As the default values of the individual variables are not summarized in any available literature, they were taken from Talhofer (2002), as well as they were obtained in a personal interview with the creators and users of databases. These default values are stated in the Table 4.14.

Before the analysis itself, it is necessary to realize that for a concrete unit of measure of database in a concrete crisis situation the value of criterion k_4 *area importance* will be constant, in this case it will equal 0.778. For this value and with the use of ideal values of all other variables, the individual functionality of this unit of measure will equal 0.7833. Together with this value it is necessary to analyze also all other influences. However, at the same time it is essential to evaluate a suitable function for the expression of level of fulfillment of criterion *database timeliness*. The Table 4.15 shows the difference of the complete individual utility value for the value of database timeliness T from Table 4.14 when using various functions:

Values "Ideal stage" are values of the individual functionality when there are ideal values of all variables except variable k_4. Values "Real stage" are then calculated for values of individual variables from Table 4.14.

In Table 4.14 apart from values of the individual variables, there are also model values of their final increases with the exception of variable k_4 for which it makes no sense to work with a change of its value. With the use of these increases it is possible to see their influence on the complete individual utility value with the use of relations from Sect. 4.3.4. The aim of this analysis is to find out which parameters in the given part of the database are suitable for modification so that the utility of the given unit of measure increased the most. That is why all increases were chosen so that the value of the appropriate criterion improved. For the evaluation itself it seems suitable to mutually compare values ΔU calculated by appropriate formulas from paragraph 4.3.4, for individual variables or for their summary within the given main criterion. The comparison may be done for instance by pair comparison with the consideration of a share of values ΔU.

Let us assume that general variables a and b will be changed at the same time. Their changes Δa and Δb evoke changes in the individual utility value ΔU_a resp.

Table 4.14 Values of the individual variables for one unit of measure of database DMU25 (all values are given in units of the individual variables)

Values of variables	α_{11}	n_{121}	n_{122}	α_{211}	α_{212}	n_{221}	n_{222}	n_{23}	n_{241}	n_{242}
	20	90	90	0	5	80	80	70	80	90
Changes of variables	$\Delta\alpha_{11}$	Δn_{121}	Δn_{122}	$\Delta\alpha_{211}$	$\Delta\alpha_{212}$	Δn_{221}	Δn_{222}	Δn_{23}	Δn_{241}	Δn_{242}
	-5	-5	-5	0	-5	5	5	5	5	5
Values of variables	n_{243}	n_{251}	n_{252}	T	k_4	k_{51}	k_{52}	α_{531}	α_{532}	α_{533}
	100	99	50	5	0778	0	0	10	10	10
Changes of variables	Δn_{243}	Δn_{251}	Δn_{252}	ΔT	Δk_4	Δk_{51}	Δk_{52}	$\Delta\alpha_{531}$	$\Delta\alpha_{532}$	$\Delta\alpha_{533}$
	5	5	5	1	0	100	100	-5	-5	-5

Table 4.15 Influence of choice of function of time dependency on the complete individual functionality

	for T in f_1	for T in f_2	for T in f_3
Ideal stage	0.7833	0.7833	0.7833
Real stage	0.4271	0.3014	0.3603

Table 4.16 Calculation of an increase of the individual utility value

U_{init}	0.4271			
	u_1	u_2	u_3	u_5
U	0.4392	0.434	0.5492	0.5336
ΔU	0.0121	0.0069	0.1221	0.1065

Table 4.17 Table of a share of changes of the individual utility value

	u_1	u_2	u_3	u_5
u_1		1.7581	0.0988	0.1133
u_2	0.5688		0.0562	0.0644
u_3	10.1218	17.7953		1.1465
u_5	8.8281	15.5209	0.8722	

U_b.The decision what is more effective to change may be made based on the value of a share of the state changes.

If $\Delta U_a / \Delta U_b > 1$, then it is more effective to improve the value of variable a as its change has a bigger impact on the complete change of the individual utility value. Otherwise it is more effective to improve the value of variable b. The following tables present the state calculations.

Table 4.16 contains values of changes of the individual utility value given by the change of variables for main criteria from Table 4.14. The default level of the individual utility value U_{init} is calculated for time function $f_1(T)$. With regards to the fact that neither space of the calculated unit of measure was changed, nor were changed the objects of interest in it, it was not necessary to calculate changes of variables in criterion k_4 area importance. It is clear from the Table 4.15 that the biggest influence on the change of individual utility value in the evaluated case lies in the database update and then standardization and data protection. This fact is documented also by results of comparison of a share of increases of the individual utility value stated in the Table 4.17.

Also the following information for workers who are supposed to make decisions what to improve in the database can be derived from the tables:

– It is more convenient to define more complex model rather than improve obeying required level of detail of data.

– Improvement of parameters evaluated by partial criteria of 1st group with criterion technical quality of database content has basically the same influence on the resulting individual utility value.
– etc.

4.4.3.1   Function Expenses

In the previous analysis, only variables that influence a change of the individual utility value were taken into consideration. Each change, however, requires expenses for which it is possible to realize such a change. Expenses may be calculated in monetary or time units as it is stated in Sect. 4.1.3. For the needs of a model example, calculation sheets for updating of one unit of measure of database DMU25 were used. These sheets are an internal document of Military Geographic and Hydrometeorological Office (MoD-GeoS, 2016).

Expenses are calculated to a unit of measure of a concrete product and according to a concrete chosen technology according to the basic calculation formula that involves expenses for:

– Direct material,
– Direct wages of production workers and control officials,
– Other direct expenses (amortization of machines, equipment, and tools, cooperation costs, costs of employee contributions, payments to the fund of cultural and social needs),
– Technical development,
– Production overheads,
– Administrative overheads.

All expenses are given in monetary units (Czech crowns, CZK) and at the same time they also are expressed as norm consumption of time on the given technological stage in real hours (RH). The real amount of expenses is calculated based on real difficulty of processing given by the complexity of terrain in a calculated unit of measure. Direct as well as indirect wages are given by financial assessment of the given profession of the individual participating employees. In the system, individual technological operations are elaborated in *technological sheets* up to the level of work operations. The technological sheets give also required qualifications for the given operation, then the norm of real hours and needed material.

Calculations of expenses are calculated for each production as well as non-production operation in the given technology according to the individual technological sheets. The individual technological operations, however, significantly differ with regards to various digital products. That is why it is good to choose a general technological scheme that may be used for calculations of expenses of any product, independently on the fact whether it is its first creation and filling, or whether it is an update of its database. It is possible to base on a general technological scheme containing the following main stages:

- Preparation and editorial work,
- Data collection, or their update,
- Saving of data into a database and its administration.

The first stage involves an establishment of an accompanying record of a unit of measure, then also collection and processing of information background including—if necessary—field geodetic work in order to provide necessary geodetic measurements. This block finishes with editorial preparation of the fulfilled content elements.

The second stage usually includes operations such as loading of the necessary data, topographic evaluation of elements or their changes, control print-outs and detailed review. Then there is the stage of stereophotogrammetric measurement and local research which are followed by finishing work, control print-out and final review before saving into the database. The stage ends with data export.

The last stage includes saving of data into the database, creation or renewal of all necessary connections and setting of appropriate database channels. This block also includes all operations concerning maintenance of databases and realization of relations between the individual users.

For the whole technology of the required product and for various classes of complexity there are *calculation sheets* elaborated. They contain a summary of expenses for the individual technological operations, or for the individual main stages. The calculation sheets respect also valid wage classes of workers. From the calculation sheets it is possible to obtain the real expenses for creation, or update of unit of measure of the database. An example of a calculation sheet is given in the Table 4.18.

4.4.3.2 Setting of an Indicator of Relative Effective Value

The final answer to the question which parameters are suitable to improve for the given unit of measure may be given also by the calculation of *the relative cost efficiency* (RCE) calculated according to formula (4.12). The following example shows the practical usage of this calculation for a unit of measure from the previous paragraphs.

The base for this calculation was a part of the current database whose complete *individual utility value* (IUV) was evaluated by variables stated in Table 4.14. The change of IUV and RCE was observed, calculated in monetary units as well as in real hours while changing chosen parameters. Change of expenses necessary for the improvement of IUV were modelled by a change of difficulty class for the given technological and work operations. The reason for this was especially the fact that the creation of time norms is rather complicated and it required comparative work in larger parts of terrain.

The Table 4.19 shows input values that were changed as well as calculation results. Only function $f_1(T)$ was used for the expression of timeliness, but it is possible to use the other functions as well. In the *first stage* only the state when there is an update of database without any change in the other characteristics was considered.

Table 4.18 Expenses for update of one unit of measure of DMU25, class of complexity 4 (out of 5)

Stage	Working operation	Norm in hours	Wages in CZK	Price of material in CZK	Social found in CZK	Total direct costs in CZK	Production overheads in CZK	Admin. expenses in CZK	Total expenses in CZK
Preliminary works	Accompanying document establishment	1	78		29	106	191	207	504
	Information sources collection	15	2328		861	3189	5727	6216	15,132
	Information sources preparation	1	155		57	213	382	414	1009
	Basic editorial preparation	12	1862		689	2552	4582	4973	12,106
	Data structure preparation	1	198		73	271	487	529	1288
Database up-dating	Topographic evaluation	179	32,606	1335	12,064	46,005	80,210	87,057	213,273
	Revision	71	13,002	1	4811	17,814	31,985	34,716	84,515
	Photogrammetric evaluation	14	2320		858	3178	5707	6195	15,080
	Revision	1	101		37	138	248	269	656
	Check-in the field preparation	3	456		169	625	1122	1217	2963

(continued)

Table 4.18 (continued)

Stage	Working operation	Norm in hours	Wages in CZK	Price of material in CZK	Social found in CZK	Total direct costs in CZK	Production overheads in CZK	Admin. expenses in CZK	Total expenses in CZK
	Check-in in the field	13	2279		843	3123	5608	6086	14,817
	Data completion	20	3647	1	1349	4998	8972	9738	23,708
	Final revision of up-dating	10	1824		675	2498	4486	4869	11,853
	Data content managing with neighbouring tiles	15	2735		1012	3747	6729	7303	17,780
	Control drawing	1	182	3	67	252	449	487	1188
	General revision	2	365		135	500	897	974	2371
Database management	Export of data	1	99		37	136	244	264	644
	Data storage in the database	1	198		73	271	487	529	1288
Total		359							420,175

Table 4.19 Calculation of changes of IUV and RCE when parameters of utility are changed

Case		1	2	3	4	5
		$T = 5$, $\alpha_{11} = 20$, $n_{23} = 70$, $n_{251} = 99$, $n_{252} = 50$	$T = 1$, $\alpha_{11} = 20$, $n_{23} = 70$, $n_{251} = 99$, $n_{252} = 50$ Difficulty class 2	$T = 1$, $\alpha_{11} = 15$, $n_{23} = 70$, $n_{251} = 99$, $n_{252} = 50$ Difficulty class 4	$T = 1$, $\alpha_{11} = 20$, $n_{23} = 70$, $n_{251} = 100$, $n_{252} = 50$ Difficulty class 3	$T = 1$, $\alpha_{11} = 20$, $n_{23} = 95$, $n_{251} = 99$, $n_{252} = 100$ Difficulty class 4	$T = 1$, $\alpha_{11} = 20$, $n_{23} = 95$, $n_{251} = 100$, $n_{252} = 100$ Difficulty class 5
IUV	0.5336		0.6860	0.6943	0.6863	0.7103	0.7106
RCE (RH)			2.6999	1.9368	2.2377	1.9814	1.6831
ΔRCE (RH)				0.7631	0.4622	0.7185	1.0168
RCE (CZK)			2.3111	1.6525	1.9118	1.6905	1.4343
ΔRCE (CZK)				0.6586	0.3993	0.6205	0.8767

The unit of measure belongs to difficulty class 2. In the *second stage* within update, the definition of content is changed (*Note* it does not concern only the stated units of measure, but the complete database of the model) and thus the number of modelled objects and phenomena increases. That is why difficulty class 4 was chosen here. The *third case* happens if the definition of content does not change but the level of fulfillment of criterion *data completeness* is improved. However, as the level of fulfillment of this criterion changes only by 1%, the difficulty class stays the same as in the first case. In the *fourth case* all missing thematic attributes are added. With regards to the fact that these thematic attributes have been fulfilled only by 50%, it is necessary to create an appropriate work capacity for their addition. That is the reason why difficulty class 4 was chosen here. In the *last case*, all missing objects and phenomena as well as their all thematic attributes are added. For this reason, difficulty class 5 was chosen here.

Table 4.19 contains the resulting values of the complete individual utility value before the update and after the update in the individual cases. Furthermore, RCE for the individual cases and changes of RCE with regards to case 1 were calculated. RCE were calculated for real hours as well as for monetary units. For clarity, the resulting value of RCE was multiplied by 1000 for monetary units, and by 1,000,000 for real hours.

For comparison reasons, the Table 4.19 may be amended also by the following graphs. From the table and graph (Fig. 4.17) it is obvious that the biggest increase of IUV occurs in case 4 and 5.

From the point of view of economic efficiency, cases 1 and 3 are the most favorable, i.e. a mere update while keeping all previous characteristics or with the addition of all missing objects and phenomena (Figs. 4.18 and 4.19).

It goes without saying that when using the calculations in the given example in which expenses in hours are accurately interlinked with employees' wages, the RCE expressed in real hours as well as monetary units will be the same. However, in case that the requirement for improvement of database functionality is decisive and it is

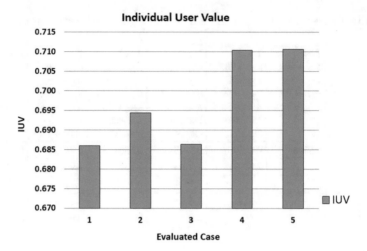

Fig. 4.17 IUV of a chosen part of database DMU25 for cases 1–5

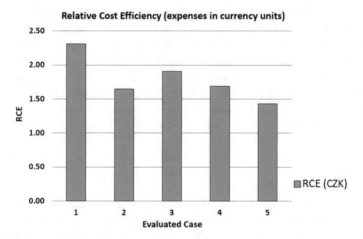

Fig. 4.18 RCE expressed in monetary units for cases 1–5

not important how many workers will cooperate on this operation, it is necessary to calculate with real hours with highest priority. Such cases may occur especially in critical situations when geoinformatics group leader has a group of geoinformatics specialists available but he/she has only a limited time for modification of the database. The provided procedure will enable him/her to optimize the decision how to meet the users' requirements.

The optimization of decisions can also be made by setting limit costs and limit functionality as mentioned in Sect. 4.1.4. Limit costs and functionality are shown in the graph (Fig. 4.20), where the individual cases under consideration are also displayed.

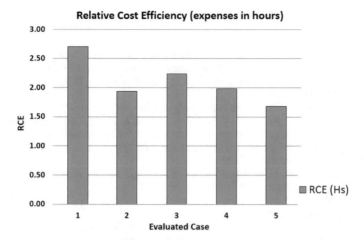

Fig. 4.19 RCE expressed in real hours for cases 1–5

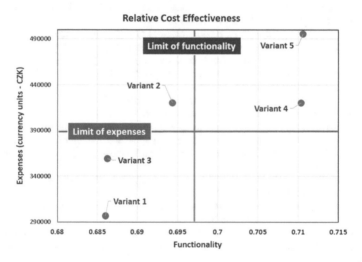

Fig. 4.20 Cost distribution and functionality of cases 1–5

It is clear from this graph this cases 1 and 3 are acceptable in terms of cost, but their functionality is insufficient. On the other hand, case 4 has a relatively high functionality but their expenses exceed the limit. From the point of user view it would the optimal to cover the expenses for this variant.

References

DoD. (1992). *Military Standard: Vector Product Format. Department of Defense*, U.S.A.
DGIWG-103. (2008). *Digital geographic information exchange standard (DIGEST)—Metadata profile of ISO19115 and ISO 19139* (2.0.0-16 December 2008 ed.). DGIWG. Retrieved from https://www.dgiwg.org/dgiwg/.
Esri. (2013). ArcGIS User documentation. Copyright © 1995–2013 Esri.
ISO. (2017). *ISO/TC 211 geographic information/geomatics*. Retrieved from International Organization for Standardization: https://www.iso.org/committee/54904/x/catalogue/.
JPL, N. (2015). *Shuttle radar topography mission*. Retrieved from NASA—Jet Propulsion Laboratory: http://www2.jpl.nasa.gov/srtm/.
Kresse, W., & Danko, D. M. (2012). *Hanbook of geographic information*. Berlin, Heidelberg: Springer-Verlag. https://doi.org/10.1007/978-3-540-72680-7.
Longley, P. A., Goodchild, M. J., Maguire, D. J., & Rhind, D. W. (2016). *Geographic information science and systems* (4th ed.). Wiley.
MGCP. (2012). *MGCP feature and attribute catalogue TRD4 v4.0.* Multinational Geospatial Co-production Program. Retrieved from https://www.mgcp.ws/web/guest.
Miklošík, F. (1987). *Over time conditionality of quality and efficiency of work in military cartography (In Czech, Časová podmíněnost kvality a efektivnosti práce ve vojenské kartografii)* (Doktoral disertation thesis ed.). Brno: Military Academy in Brno.
Miklošík, F., & Rybansky, M. (1997). Ragular changes of topographic maps content (In Czech, Zákonité změny aktuálnosti obsahu topografických map). *Kartografické listy, 1997*(5), 5–12.
Miles, L. D. (1989). *Techniques of value analysis engeneering* (3rd ed.). USA: Eleanor Miles Walker.
MoD-GeoS. (2006). *Catalogue of topographic objects DMU100*. Dobruska: Ministry of Defence of the Czech Republic, Geographic Service.
MoD-GeoS. (2013). *Catalogue of the Topographic Objects DMU25* (7.3 ed.). Dobruska: Ministry of Defence of the Czech Republic, Geographic Service.
MoD-GeoS. (2016). Cost calculation for planning, evaluation, pricing and analysis calculations (In Czech). *Internal regulation*. Dobruska: Military geographic and Hydrometeorological Institute.
OGC. (2016). *Keyhole Markup Language*. Retrieved from Keyhole Markup Language: https://dev elopers.google.com/kml/.
STANAG 2215. (2010). *Evaluation of land maps, aeronautical charts and digital topographic data* (7 ed.). NATO Standardization Agency.
Talhofer, V. (2002). *Possibilitis of improvement of digital geographic data (In Czech)* (Habilitation thesis ed.). Brno: Military academy in Brno.
Talhofer, V. (2004). Digital Geographic Data: Potential Evaluation. *AGILE 2004, 7th Conference on Geographic Information Science, Conference proceedings* (pp. 675—686). Heraclion, Crete, Greece: AGILE.
Talhofer, V., & Hofmann, A. (2009). Possibilities of evaluation of digital geographic data quality and reliability. *24th International Cartographic Conference, The World's Geo-Spatial Solutions* (p. 11). Santiago de Chilie: ICA/ACI.
Talhofer, V., Hoskova-Mayerova, S., & Hofmann, A. (2012). Improvement of digital geographic data quality. *International Journal of Production Research, 50*(17), 4846–4859.

Chapter 5
Multi-criterial Analysis of Geographic Spatial Conditions

Multivariate techniques are quite frequent and complex. They have mathematical basis in multi-criterial analysis for decision-making processes.

5.1 Principle of Multi-criterial Analysis

A decision-making analysis is a set of systematically performed procedures for analysing complex decision-making tasks. The basic strategy of solution consists in dividing tasks into smaller and more comprehensible parts in a logical way so that a meaningful solution is found. The types of decision-making tasks that geographers and "space designers" are interested in include a set of feasible alternatives as well as also contradictory and incommensurable evaluation criteria. The alternatives are usually evaluated by more than one person. Evaluators usually have various priorities; resulting on them, all possible alternatives of decisions are evaluated (Fiala, Jablonsky, & Manas, 1997).

Conventional techniques of multi-criterial decision-making are considered to be non-spatial in terms they assume space homogeneity within the researched area. This assumption is not realistic for many cases as evaluation criteria change in space.

Therefore, the spatial multi-criterial decision-making analysis represents a significant exception to such a conventional multi-criterial decision-making analysis due to the presence of a spatial component; it requires data corresponding to the evaluation criteria as well as geographic position of alternatives. Data are processed by using GIS and at the same time using the multi-criterial decision-making method. Spatial multi-criterial decision-making analysis may be considered as a process combining and modifying geographic data into the resulting decision. Decision rules define the relationship between the input and output map.

© Springer International Publishing AG, part of Springer Nature 2019
V. Talhofer et al., *Quality of Spatial Data in Command and Control System*, Studies in Systems, Decision and Control 168, https://doi.org/10.1007/978-3-319-94562-0_5

GIS and multi-criterial decision-making analysis may benefit from each other. On the one hand, techniques and procedures in GIS play an important role in analysing multi-criterial decision-making problem as they provide unique tools for automation, control and analysis of spatial data. On the other hand, multi-criterial decision-making and a broad scale of related methodologies offer a wide set of techniques and procedures for revealing priorities of the person involved in making the decision.

In terms of knowledge and skills of a potential user, the methods of multi-criterial analysis are more demanding comparing to the method of key factors, particularly in the sense of setting relative significance of the individual factors (indicators). Their advantage consists in the easy repeatability (e.g., in the following updates); of course, under assumption that no significant changes of situation have occurred within the researched area.

When working with individual data, the choice of a suitable method of multi-criterial analysis and setting relative significance of area indicators is another way how to cover characteristics of various areas.

However, the setting of relative significance is usually quite difficult; that is why methods (e.g., Saaty's or Fuller's methods) constructing these estimations based on simpler subjective information are used. It follows the principle: the higher importance, the bigger its priority is.

Spatial multi-criterial analysis may be defined as a set of techniques serving for an analysis of geographic phenomena where results of the analysis depend on spatial arrangement of the given phenomena. At the same time, it is a part of widely defined spatial analysis. Despite the fact that most spatial decision-making problems is in its principle multi-criterial, the process of multi-criterial decision-making is not well established or effectively integrated in spatial analyses and GIS (Malczewski, 1999).

5.2   GIS and Support of Decision-Making

One of the aims of GIS is to provide support in a decision-making process. The abilities of GIS in support of spatial decision-making may be analysed in the context of a decision-making procedure. The most widely spread and accepted generalization of decision-making procedure consists of three basic parts (Malczewski, 1999; Brozova, Houska, & Subrt, 2003).

Intelligence

At this stage, the problem is investigated, analysed and described. Each decision-making process begins with recognition of a task. In that step, rough data are acquired, processed and researched. It is researched whether they provide any clue that may help with identification of new tasks or possibilities for improvement.

Design

At this stage, proposing and analysing possible solutions are carried out. There is specified alternatives (possibilities) to solve. The suggestion involves the development and analysis of a set of possible solutions. This point uses a model in order to help the person performing the analysis with defining the set of possible alternatives of the solution. The model is a simplified representation of reality. It represents a decision-making situation with the help of structuring and forming available data and information about the task. Possible alternatives for making a decision are deduced using the manipulation and data analysis saved in GIS.

Choice

The next stage consists in selecting possible alternatives, i.e., specifying the most relevant decision. While creating possible alternatives is just a matter of design step, evaluating the alternatives is the main step of the phase Choice. It involves choosing a certain alternative from a set of feasible alternatives. In this phase, each alternative is evaluated and analysed in relation to other alternatives within a specific decision-making rule. The rules are used to rank the alternatives that come into consideration.

The three described stages in a decision-making process do not always have to come one by one. In each point of decision-making, it may be necessary to come back to the previous stage. Each level of a decision-making process requires different kinds of information. In the context of spatial decision-making analysis, the key question is how and in what extent GIS may provide support during a decision-making process

5.3 Elements of Multi-criterial Decision-Making Analysis

Tasks of multi-criterial decision-making analysis generally consist of six elements:

– Goal or set of goals that we want the achieve,
– Person or a group of persons that is involved in the decision-making process with their priorities during a decision-making process that relate to evaluation criteria,
– A set of evaluation criteria, on the basis of which the decision-maker evaluates the alternative ways of procedure,
– A set of alternatives of the decision,
– A set of uncontrollable variables or condition of the surroundings in which the decision-making task takes place,
– A set of results or consequences related to each alternative of the decision.

All decisions are made in a certain surrounding context that is why they also include many factors that are out of sight of the person who performs the analysis. These factors are presented as conditions of surroundings. A condition of surroundings, for instance, may be the state of economy (recession, inflation) or weather conditions.

5.4   Framework of Spatial Multi-criterial Decision-Making Analysis

Decision-making is a process. It involves a lot of actions starting from task recognition and definition, ending with a recommendation. It has been proved that the quality of a decision-making process depends on the order, in which the actions are performed (Malczewski, 1999).

Task Recognition and Definition

All decision-making tasks begin with a recognition and task definition. The content of the task may be defined as a difference between the current and required state of things. Defining the task overlays the phase of Intelligence mentioned in the previous text.

Evaluation Criteria

Criterion is a judgment or a rule that serves to test the suitability of alternatives of a decision. In time when the task has been defined, establishment of evaluation criteria comes next.

This step involves finding all goals that describe all parts relating to the decision-making task and finding all means to achieve these goals. These means are called attributes. Proportional scales for all attributes should be set in this step as well. Evaluation criteria are linked with geographic units and relations between them and that is the reason why they may be projected in the form of maps. There are two kinds of criterial maps. A map of evaluation criteria and a map of restriction. Maps of evaluation criteria project a certain attribute that may be used for the evaluation of an alternative. These maps are also presented as maps of attributes (e.g. thematic maps). A map of restriction projects a restriction of attribute values. Each criterion should be complex, measurable, unambiguous and comprehensible.

For spatial multi-criterial analysis, each criterion should be projected with the help of map layers. The analysis requires the values contained in the individual layers to be transformable into unified units. An example of the simplest transformation of gross data into standardized units is a linear scale transformation. Each value of a set of values is divided by the highest value of the set. The higher the standardized value, the more positive criterion value.

Alternatives

As already mentioned, the course of finding alternatives should be related to the set of evaluation criteria. The basic rule how alternatives should be created is to achieve evaluation criteria of the decision-making task as well as possible. In a raster GIS, the alternatives are represented by individual raster layers, and in vector GIS the attributes may be represented by points, lines or polygons.

In the real world, there are only a few decision-making tasks where alternatives may be considered as unrestricted. If there are set restrictions for alternatives, then alternatives which obey such restrictions are considered to be feasible. In case of GIS, the restrictions are used for exclusion of unsuitable points, lines, polygons or raster layers characterized with the help of a given unsuitable attribute.

Criterion Weights

Priorities of the person who makes the decision are mostly expressed with the help of weights that describe the level of importance for criteria. One of the methods how to assign weights is ranking criteria according to importance. Weights must be normalized in such a way that their sum equals 1. Criterion weights may be calculated with the help of ranking or evaluation of criteria. Defining and assigning weights is the main step for setting of priorities. Due to the fact that each person or a group of persons involved in the process of decision-making has specific priorities, it is logical that the results of an analysis will always be unique.

Decision Rules

This step uses and follows the results of the previous three steps. One-dimensional measurement (geographic data) and evaluation (priority and inaccuracy) must be unified in order to obtain the total evaluation of alternatives. This is done with the help of a suitable decision rule or unification function. Decision rules order how to rank alternatives from the most suitable to the least suitable. For spatial multi-criterial analysis, Simple Additive Weighting belongs among the best and most frequently used methods. It is also known as weighted linear combination or scoring methods. It is a simple and most often used multi-attribute decision technique based on the weighted average. An evaluation score is calculated for each alternative by multiplying the scaled value given to the alternative of that attribute with the weights of relative importance directly assigned by decision maker followed by summing of the products for all criteria (Afshari, Mojahed, & Rosnah, 2010).

The advantage of this method is that it is a proportional linear transformation of the raw data, which means that the relative order of magnitude of the standardized scores remains equal.

As the criterion is represented with the help of map layers, for summing the method of overlay is used, in the language of map algebra it is function LocalSum.

Sensitivity Analysis

The subsequent step in the whole procedure is a sensitivity analysis whose aim is to define the robustness of the whole project. Sensitivity analysis defines whether the recommended alternative of decision is sensitive to changes in the input data. Its goal is to identify the impact of a change in input data (geographic data or priority of those performing the analysis) and what changes they cause in the result, recommendation and consequences. If the changes do not have a significant impact on the analysis result, then the ranking of alternatives is considered as robust. If it is not the case, it is possible to use newly obtained information and come back a few steps. Sensitivity analysis may help the researcher to better understand the task and the individual parts and thus contribute to the choice of the best alternative of the decision.

Recommendation

The last step of the analysis is a recommendation for future actions that should be based on the previous steps. The recommendation is usually complemented by an analysis of potential consequences of the given decision as well.

Methods for Verification of Multi-criterial Analysis

MCA research on the validity issue has focused on the problem of selecting an MCA method under various decision contexts along two lines of development:

– experimental comparisons of MCA methods for examining their appropriateness of use and/or theoretical validity (Zanakis, Solomon, Wishart, & Dublish, 1998), and
– method selection procedures for specific characteristics of the decision problem and distinct features of available methods in the form of decision support systems (Ozernoy, 1992) or as general selection principles (Yeh & Willis, 2001).

While the results of experimental comparisons cannot be used as guidelines for a decision-making to select a proper multi-analyses method for an application (Ozernoy, 1992), the method selection procedures may not always make a clear unequivocal choice (Yeh & Willis, 2001), in particular between methods of the same category. Due to their implicit and explicit assumptions, the applicability of the methods selected remains uncertain, as evidenced by the fact that these selection procedures do not normally examine the validity of decision outcomes. Despite the significant development in multi-analyses method selection research, the validity of decision out-comes remains an open issue. This is mainly due to the fact that the "True" cardinal ranking of alternatives is not known.

5.5 Methodology of Solution of Geo-Spatial Data Homogenization, the Use of Fuzzy Sets

Fuzzy set theory is an extension of classical set theory that "Allows solving a lot of problems related to dealing the imprecise and uncertain data" (Balmat, Lafont, Maifret, & Pessel, 2011). It has many advantages. Fuzzy logic "Takes into account the insufficient information and the evolution of available knowledge" (Balmat, Lafont, Maifret, & Pessel, 2011). It allows imprecise input. It allows for a few rules to encompass problems with great complexity. For disadvantages, fuzzy systems can sometimes be difficult to develop. In many cases, they can require numerous simulations before being able to be used in the real world (Velasquez & Hester, 2013).

Fuzzy set theory is established and has been used in applications such as engineering, economic, environmental, social, medical, and management. Many of these types of problems take advantage of the availability of imprecise input. These types of applications favor a method that embraces vagueness and can be tested numerous times before real-world application. Application of fuzzy sets in the field of decision-making, for the majority parts, is based on "fuzzifications" of classical decision analysis methodology. There are many ways to introduce fuzziness into the existing models of decision-making. A generic models for fuzzy decision was suggested by Bellman and Zadeh (1970). This model was created using analogy to the classical decision problem. The main task it to "Optimize a criterion function subject to a set of constrains". It is due to the fact that the fuzzy criterion is characterized by its membership function, and so they are constrains. The fuzzy criteria and fuzzy constrains are combined to form a decision. The relationship between criteria and constrains is in fuzzy decision symmetric in the sense that the two are treated operationally in the same way. Both (criteria and constrains) are defined as a subset of the decision space. Using the upper mention model of Bellman and Zadeh, the best alternative is that which has the highest grade of membership in the intersection of criteria and constrains.

For more, detailed information concerning this topic the reader is referred to study the book (Malczewski, 1999).

References

Afshari, A., Mojahed, M., & Rosnah, Y. M. (2010). Simple additive weighting approach to personnel selection problem. *International Journal of Innovation, Management and Technology, 1*(5).

Balmat, J., Lafont, F., Maifret, R., & Pessel, N. (2011). A decision-making system to maritime risk assessment. *Ocean Engineering, 38*(1), 171–176.

Bellman, R. E., & Zadeh, L. A. (1970). Decision making in a fuzzy environment. *Management Science, 17,* 141–164. https://doi.org/10.1287/mnsc.17.4.B141.

Brozova, H., Houska, M., & Subrt, T. (2003). *Models for multicriteria decision making (in Czech)*. Prague: Czech University of Life Sciences.

Fiala, P., Jablonsky, J., & Manas, M. (1997). *Multicriteria decision making (in Czech)*. Prague: University of Economics, Prague.

Malczewski, J. (1999). *GIS and multicriteria decision analysis*. New York: Wiley.

Ozernoy, V. M. (1992). INFOR: Choosing the "best" multiple criteria decision-making method. *Information Systems and Operational Research, 30*(2), 159–171.

Velasquez, M., & Hester, P. T. (2013). An analysis of multi-criteria decision making methods. *International Journal of Operations Research, 10*(2), 56–66.

Yeh, C.-H., & Willis, R. J. (2001). A validation procedure for multicriteria analysis: Application to the selection of scholarship student. *Asia Pacific Management Review, 6*(1), 39–52.

Zanakis, S., Solomon, A., Wishart, N., & Dublish, S. (1998). Multi-attribute decision making: A simulation comparison of select methods. *European Journal of Operational Research, 107,* 507–519.

Chapter 6
The Influence of Quality of Geospatial Data on the Decision-Making Processes

As it was mentioned in Chap. 1, the strength of digital geographic data together with tools for working with them lies in the possibility to create a spatial geographical analysis in which the influences of geographical environment on the intended or realized activity are evaluated (Prochazka, Rapcan, & Hubacek, 2002).

According to Udvorka (2006) spatial analyses represent a collection of techniques that were created in various fields and whose aim was data analysis with an emphasis on their spatial relations. These analyses are a set of techniques for analysis and modelling of localized objects where results of the analysis depend on spatial arrangement of an object and their characteristics. The goals of spatial analyses vary according to the fields of application and it is difficult to find a universal division. Nevertheless, on a certain level of generality it is possible to define the following general categories of goals of spatial analyses:

- Description of objects, or events in the researched space (including description of their arrangement),
- Choice of a certain place based on fulfillment of a certain set of conditions (or generally according to a certain decision-making scheme) or research of the extent of fulfillment of certain conditions in a certain place or area,
- Interpretation of processes that lead to the observed state of arrangement of objects or events in the researched space (systematic research), e.g. interpretation of occurrence of an event, explanation of development of an area in time, etc.,
- Optimization of arrangement of objects or phenomena in the researched area with the help of e.g. localization and allocation tasks, choice how to distribute the flow (distribution of the employed, children in schools, goods), but also e.g. suggestion of a suitable system of sampling,
- Improvement of the ability to predict and control objects or events in the researched space (using predictive models),
- Reduction of original amount of data into a smaller, more economical and well-arranged set of data.

© Springer International Publishing AG, part of Springer Nature 2019
V. Talhofer et al., *Quality of Spatial Data in Command and Control System*, Studies in Systems, Decision and Control 168, https://doi.org/10.1007/978-3-319-94562-0_6

The provided overview of goals of spatial analysis is not and even cannot be complete as with the development of geo-information technologies there are new and new forms of application of spatial analyses and thus also new goals.

The goals of spatial analyses are reached by various methods of analysis. It is possible to approach the division of the methods of spatial analyses from many different points of view, e.g. from the point of used procedures, ways of processing, level of processing, number of simultaneously studied phenomena, or according to the type of presentation of spatial objects of phenomena.

The methods are usually divided according to:

- Used procedures (applicable techniques),
- Ways of data processing,
- Type of spatial representation.

According to the *used procedures* it is possible to divide the methods to:

- Statistical spatial analyses of data (spatial statistics) that are closely related to mathematical statistics and they are realized e.g. in map analysis or algebra,
- Methods of mathematical modelling—e.g. creation and analyses of multivariate or regression models,
- Interpolation methods,
- Localization and allocation methods,
- Network analyses where the subjected network is traffic, hydrological or engineering networks,
- Other analyses of surrounding and connection—e.g. image processing techniques used for obtaining geometric characteristics of images or textures, gravitational analyses, etc.

According to the *ways of data processing*, the methods may be divided to:

- Visualization,
- Exploration,
- Modelling.

Data processing may be understood as a certain sequence of steps creating certain phases or levels of data processing. The first step in spatial data analysis is usually visualization of primary or in a basic way adjusted data. Further phases, i.e. exploratory or modelling, may or do not have to be realized, but if they are performed, it is always after the basic visualization phase. Results of these phases may be (and usually are) presented with the help of visualization methods. Thus, it is possible to further divide visualization methods according to their use before or after data processing to preparatory (pre-modelling) and output (post-modelling) visualization methods.

Visualization methods focus on the projection of spatial data without a modification of graphic component of data. They do not require statistical data processing, maybe only while definition of class borders. Creating of maps, choropleth maps or diagram maps is often used with imaging methods. The resulting compositions document objects and phenomena in the researched area and they are visually interpreted.

Exploratory methods do not project the original data, but they use modified spatial data. They use such forms of data processing as e.g. smoothing, transformation, filtration, etc. They often perform summarization of values. They deal with research of map texture, relations and detection of anomalies.

Imaging and exploratory methods are used in *data analysis*, whose goal is basic research of data characteristics. Among the main features of data analysis, we can mention identification of data required for purposes of detection of data texture, forming of hypothesis based on registered data and for some aspects an evaluation of a model (e.g. the credibility of a model). It is recommended to use simple, intuitive and statistically robust methods.

Modelling methods represent the third group of methods of spatial analysis. They are based on the creation of a suitable model, verification of its suitability for the given purpose (e.g. explanation of influence of certain factors), subsequently the use of parameter of a model for interpretation of a phenomenon, relation or the use of a model for research of the consequence of certain changes in parameters, time development, etc. In the field of modelling, predictive modelling using localization and allocation tasks belong to the most important tasks.

According to the *type of spatial representation* the methods may be divided to methods suitable for continuous representation and for discreet representation that may be further divided according to the type of geometric primitives used for representation.

Statistical spatial analyses include methods based on stochastic (random) nature of arrangement and relation. They offer quite wide range of applications. Sometimes they are referred to as spatial statistics. According to the number of simultaneously researched characteristics, it is possible to divide the statistical methods into *monovariate* (one-dimensional) and *multivariate* (analyses of multidimensional objects/events).

Monovariate statistical analyses work simultaneously only with one characteristics of an object, while multivariate statistics use more characteristics at the same time. With multivariate techniques we may further specify whether they study more characteristics simultaneously only visually (and then analyses and interpretation is done visually as well as mentally), or whether these are really multivariate methods. The division according to the nature of statistical techniques uses the analogy of the traditional division of statistical techniques to descriptive (typically calculation of statistical characteristics) and inductive (based on the study of choice we consider the characteristics of the whole, carry out testing of hypotheses). According to this principle, techniques of statistical spatial analysis are divided to:

– Descriptive (centrographic techniques)—first of all quantitative measurement of characteristics of position and characteristics of distraction,
– Inferential (analyses of texture)—they define whether distribution is or is not random, describe relations between two or more quantities.

Distribution of geo-element is understood as a texture (pattern) that geo-elements create by their distribution in the researched part of area. From the point of view of geo-element texture, there are three basic types of distribution recognized. A more detailed overview of basic techniques for performance of statistical spatial analyses includes:

– Simple descriptive analyses, data transformation and summarization,
– Methods of nearest neighbour and K-function,
– Quadrant, core and Bayes smoothing methods,
– Spatial autocorrelation and covariance structures,
– Geostatistical and spatial econometric modelling,
– Spatial generalized linear modelling,
– Multivariate techniques,
– Spatial interaction models.

References

Prochazka, D., Rapcan, V., & Hubacek, M. (2002). Using ModSAF in Czech Army: The current status. In *6th World Multi-Conference on Systemics, Cybernetics and Informatics (SCI 2002)/8th International Conference on Information Systems Analysis and Synthesis (ISAS 2002)* (pp. 344–348). Orlando, FL: INT INST INFORMATICS & SYSTEMICS.

Udvorka, P. (2006). *Map alegbra and its usage in the geographic analysis (In Czech)* (Dissertation Thesis ed.). Brno: University of Defence.

Chapter 7
The Use of MCA in Decision-Making Processes on an Example of Terrain Passability Model

General approaches of MCA may be used in many areas of decision-making processes. Chapter 4 showed its applications for data models of digital geoinformation and the evaluation of their functionality. The following text is dedicated to the use of MCA for solutions of geo-process models. To illustrate the complexity of the method of solution, an example of modelling of passing through terrain with off-road vehicles was chosen.

7.1 Modelling of Geographic Tasks

For modelling of geographic problems within geo-information technologies especially geo-process models are used. As it was mentioned in Chap. 2, *geo-process models* capture interactions of objects that are saved in a representative model. However, there might be also external data and information entering the models, which usually specify the default conditions in which geo-process models work.

Geo-process models are mostly created based on an analysis of behavior of objects within the given problem. Geo-process models may be used to describe processes, often also for prediction of what will happen when a certain action takes place.

Each analytical tool of GIS may be considered to be a geo-process model. Some geo-process models are simple, others more complicated. Complex models may be created by including logic and combination of more geo-process models with the use of so-called map algebra.

© Springer International Publishing AG, part of Springer Nature 2019   133
V. Talhofer et al., *Quality of Spatial Data in Command and Control System*, Studies
in Systems, Decision and Control 168, https://doi.org/10.1007/978-3-319-94562-0_7

Fig. 7.1 Adding of raster layers

Fig. 7.2 Combination of raster layers with the use of truth logic

7.1.1 Theoretical Basics for the Creation of Spatial Analyses

One of the basic geo-process models is e.g. adding two raster layers (Fig. 7.1).

A more complex model may use the added logic as seen in the picture (Fig. 7.2). Here there are two raster layers combined and with the help of truth logic expressed by Boolean algebra sandy and at the same time dry soil is searched for. The result is a new raster layer that contains only values T (true) and F (false).

A complex model is created by adding special tools with an algorithm generating analytical result and combining even rather complicated procedures, e.g. calculation of non-Euclidean distance or hydrological dispersion of pollution of groundwater (Fig. 7.3).

From the elevation model DMR5 the mistakes are removed. From the repaired model FillSink the layer of water stream flow is calculated using tool FlowDirection. From obtained layer Flow the accumulation of water is calculated using FlowAccum. The raster set Accum is the result of this step. The last step is selection of pixels which create the net of water streams.

Complicated complex models combine multiple tools and logic (Fig. 7.4).

Several types of maps can be created from DMR5. The map of slopes will be used for overlay operation, reclassified map of slopes is called ReclSlo. ReclObs is the reclassified map of visibility from 4 different positions. Resulting CostMap gives information about passability with respect to slope and visibility.

The geo-process model for capturing of reality necessary for a solution of a given problem should be as simple as possible. Sometimes only a single operation or tool is enough, in other cases it is necessary to create a complicated model containing even hundreds of tools.

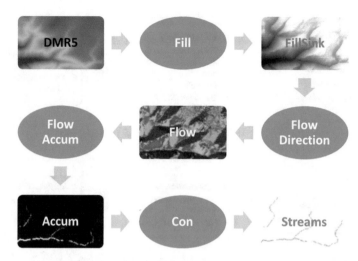

Fig. 7.3 Complex model

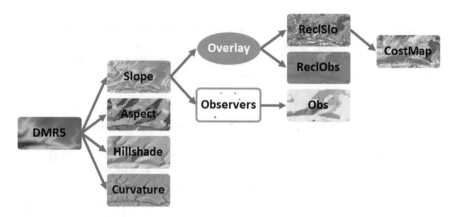

Fig. 7.4 Complex model with logic

There are many kinds of geo-process models for solutions of various problems. The problems may be put in groups solving similar problems. One of the possible divisions is provided by Mitchell (1999):

- *Modelling of suitable localities*, where an example can be searching for a suitable space for new construction or for optimization of distribution of observation sensors in countryside,
- *Modelling of distance*, e.g. setting of evacuation zones in case of emergency,
- *Hydrological modelling* which looks for drainage conditions of water from the surface in the researched area,
- *Modelling of functional surfaces*, e.g. modelling of the level of environmental pollution in various places in a given area.

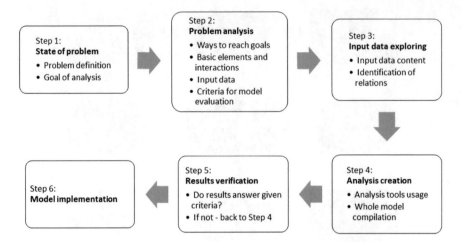

Fig. 7.5 Conceptual model of geo-process model—modified by Mitchell (1999)

Each geo-process model, however, must follow a detailed analysis of functional relations between the individual variables that enter the model as well as data streams that secure the realization of the model. Especially with complex process models it is necessary to prepare a conceptual model of the solution beforehand.

7.1.1.1 Conceptual Model of Solution of a Geographic Problem

Modelling of geographic problems based on geographic analyses follows the general principles of MCA, modelling of analytical processes and methodology of their solutions. The whole process must be commenced with a detailed analysis of the given task and by setting the basic direction in which the analysis will go. It means to create *a conceptual model of solution of geographic analysis.* The essential steps of the conceptual model may be described by the following structure (Mitchell, 1999):

– Problem definition,
– Goal that shall be reached,
– Problem analysis,
– Research of input data files,
– Performance of analysis,
– Verification of results of modelling,
– Implementation of results.

The given structure of conceptual model of geo-process model may be illustrated in the block diagram (Fig. 7.5).

Within the definition of the problem there are basic tasks set which need to be solved. The initial problem definition is often quite vague. For the creation of the conceptual model of its solution it is necessary to further specify the problem, i.e. to set exactly what will be solved in the process of modelling. After such specification it is possible to approach to solution of step 1, it means to define the goal or goals of analysis including the expected reliability of final results of the analysis.

If the goals of analysis are clear, it is possible within step 2 to approach to the detailed analysis of the whole problem. It means to divide it into a set of tasks, to identify basic elements or factors and their interactions that are necessary for reaching the set goals. Also, it is necessary to discover functional relations between the individual factors, i.e. to define which factors are mutually independent and which are dependent on each other. Furthermore, it is necessary to set criteria which will be used for evaluation of all semi-results as well as final results, including the definition of weight of the individual criteria.

Following step 3 is dedicated to research of input data files. It is important to understand how spatial and attributive characteristics of the individual objects and phenomena in countryside and relations between them are modelled, i.e. to understand correctly the representative data model. To understand such relation, it is necessary to thoroughly research the input data and their characteristics. If there are more usable variants of input data, it is possible to work with their functionality and proportional value. Thus, it is possible to optimize the complete solution from the point of view of reaching the required goals of analysis, as well as from the point of view of composition of data sets, expenses for their acquisition, updating, etc.

The whole process model of analysis is created in step 4. For the creation it is possible to use inbuilt tools of the used program system GIS as well as tools created e.g. in a suitable program language. It is possible to compile a simple process model from simple inbuilt tools, e.g. tools for selections, tools of map algebra, etc. More complex process models are created according to principles of creation of complex programs including a detailed description of data streams of input and temporary data.

After finished compilation of tools used to perform an analysis and verification of its basic functionality it is necessary to verify results of modelling with the help of tests over real data. The model should be tested using the input data and parameters corresponding to the researched area with a possible change of some parameters in order to reach a better result. If several models are created, it is purposeful to determine which of them best and unambiguously complies with the original goals. Obtained results are then compared with the set criteria and the level of fulfillment of these criteria is evaluated. If the model does not fulfill the criteria, or fulfills them only partially, it is necessary to come back to the previous step and modify the model so that the inputs improved.

Once the problem is conceptually solved and it is verified that the results meet the original expectations and comply with the set criteria, it is possible to realize the result, i.e. to implement the model into practice, either as a separate solution or a solution of partial analysis within e.g. the system of command and control, decision-making systems, etc.

For concretization of the solution of conceptual model, it is possible in systems of command and control to use also established or prepared scenarios of solutions of typical tasks that in crisis management are often a part of standard operational procedures. With the help of such scenarios it is possible to identify most goals, used work procedures to reach them, essential input data, system of cooperation of interested units, etc., which are described in the first three steps in the block diagram of conceptual model (Fig. 7.5). Steps 4–6, however, need to be solved for the given scenario.

The application of the given conceptual model on a simple multi-criterial analysis is the subject of the following paragraph.

7.1.1.2   Application of Multi-criterial Analysis for Solution of a Concrete Geographic Problem

This paragraph describes a model example of searching for a suitable locality for deployment of logistics centre for units in action when solving a crisis of large extent. The given example resembles a real situation; however, it is very simplified so that it was obvious how the whole conceptual model is created and verified.

For many activities when solving crisis situations in terrain it is necessary to have logistics facilities for units in action available. So, the geographic problem is to identify suitable places in the given area where it would be possible to deploy such logistics facilities so that in the given space the facilities were functional and configuration of terrain did not restrict it or restricted them only to a reasonable extent.

The goal of the solution is to find places that meet the conditions for the given space. These conditions might be e.g. minimal size of the area for deployment of elements of logistics facilities, maximum inclination of slopes, corresponding distance from communications, waters, places of solution of emergencies, etc.

Based on the set goal it is possible to move to the second step—*problem analysis*, i.e. to commence with searching for ways how to reach it. First of all, it is necessary to understand factors that influence the whole solution of a problem and to organize these factors into a clear structure. For the individual factors it is then necessary to choose the evaluation criteria and set their weights. In order to do this, the whole problem must be analyzed in detail.

In the given case, the choice of a suitable place is influenced first of all by the factor of terrain relief, its surface, communication relations, vegetation cover, hydrological relations and the distribution of crisis situations that are the subject of solution.

In the mentioned model case it is required that the chosen space was in a relatively flat terrain, on a suitable soil that would even after some rain still have sufficient bearing capacity, within reach of communications, in treeless countryside and in safe distance from water courses.

The individual factors are evaluated by measurable or classifiable criteria. The following criteria may be used in the mentioned example (Table 7.1).

Furthermore, interactions between the individual factors are looked for. In this simplified example, however, it is not necessary to evaluate mutual relations between the individual factors as the individual factors do not depend on each other and their

Table 7.1 Evaluation criteria for the choice of a suitable place for dislocation of logistics facilities

Factor	Evaluation criterion
Relief of terrain	Inclination of slopes
Communication relations	Distance from the nearest communication
Waters	Distance from the nearest water course and body of water
Soil type	Land Use Code, alternatively Vehicle Cone Index (VCI) or classification of soil types
Vegetation	Size of treeless area

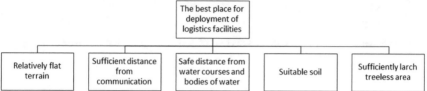

Fig. 7.6 Hierarchy of conditions of process scheme

influence on the problem solution may be determined separately. Then the whole hierarchy of condition of process may be expressed for instance in the form of hierarchical scheme (Fig. 7.6).

In the next phase, suitable input data that enable to set the influence of the individual factors with the help of evaluation criteria are looked for. As the basic data layers may be in vector as well as raster format that contain needed data in various data types, it is necessary to identify how these data will be used for the individual evaluation criteria. For evaluation criteria the minimal values of criteria for positive evaluation of the given factor in the whole model of solution of the given problem are established.

Within the solution of process scheme of the model example, it is possible to solve the individual factors one by one.

The factor of terrain relief resembles in the requirement for relatively small inclination of relief. In order to find such localities, it is necessary to create a map of inclination of relief. The process model will then contain a calculation of inclination of slopes where the input data layer will be a digital elevation model.

The second factor deals with the distance of communications from the potential places of deployment of logistics center. It is important for the logistics center to be—if possible—within the reach of communication network and vehicles move on unpaved surface as little as possible. That is why it is necessary to create the map of distances to the nearest communication. That is the reason why the model will contain the process of distance calculation for which the communication network will be the input data layer.

The third factor deals with a safe distance from water sources. The location of the logistics center brings the danger of pollution of waters with oil and other dangerous

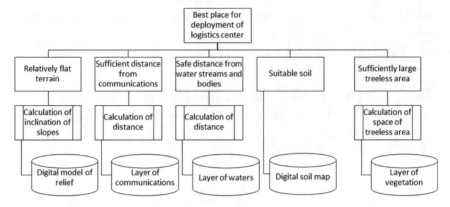

Fig. 7.7 Scheme of process model of selection of a suitable place for deployment of logistics facilities

substances. That is the reason why the next step is to create the map of distance from surface water sources. The model will contain the process of calculation of distances to water sources for which the layer of waters will be the input data layer.

The fourth factor considers the soil type. It is necessary to decide what soil from the point of view of soil type is suitable for an activity in the given area. To a certain extent it is a subjective process that depends on the solution of a problem and conditions of movement in the given locality. It seems most suitable to set a hierarchy from the most to the least suitable. For the assessment it is possible to use alternatively Land Use Code (LUC) or standardized soil type codes that may possibly be connected with values of Vehicle Cone Index (VCI) for decisive mobile equipment that will move in the given area. Digital soil map containing LUC or soil types possibly complemented with codes VCI will be the input data.

The last factor deals with the influence of vegetation, especially forest covers. For deployment of logistics center, it is ideal to find—if possible—continuous treeless area with given minimal dimensions and space. A map of treeless areas with their size will be created in this step. So the model must contain a process of searching of treeless areas. The layer of vegetation with be used as the input data.

Based on the given analysis of facts and criteria of their evaluation, it is possible to create a process scheme that contains basic conditions, as well as input data, processes and outcomes, as seen in Fig. 7.7.

In order to consider that some factors are more significant for the final solution than others, it is necessary to set weights of the individual factors. For the model example, setting of percent significance and normalized weights of the evaluation criteria are given in Table 7.2.

Once the process scheme of solution is created, the second step of conceptual model is finished and it is possible to move on to research of input data, i.e. the third step.

Table 7.2 Weights of factors and evaluation criteria

Factor	Evaluation criterion	Impact of factor in %	Criterion weight—normalized value
Relief of terrain	Inclination of slope	13	0.13
Communication relations	Distance from the nearest communication	42	0.42
Waters	Distance from the nearest water course	21	0.21
Soil type	Code of soil use or code of soil type	8	0.08
Vegetation	Size of treeless area	16	0.16

Research of input data involves a detailed analysis of a structure and content of data sources that are available. It is necessary to thoroughly evaluate their utility characteristics, especially the content of data files, structure of their attributes, their geometric and thematic level of detail. Furthermore, it is also necessary to evaluate how the data will be handled in the individual partial processes, what tools for handling with them will be used, how the data conform to these tools. This question is closely related to the data format and a possible necessity of transformation of the original data into such data format that is suitable for the used tools.

The model example used DMU25 as the positional model, elevation model DMR4 and soil database (SD) (Novák, 2000). The whole analysis was performed on an area east of city Brno, Czech Republic, whose dimensions were 13 × 13 km.

The phase of *analyzing* deals with the selection of suitable tools for partial analyses of influence of the individual factors and for evaluation of the total influence of all factors. Furthermore, it deals with classification of obtained information so that it was possible to classify this information in a unified evaluation scale.

With the help of suitable tools, entry layers for the analysis are created, as shown in the process scheme in the picture (Fig. 7.7). For the calculation of inclination of slopes, tool Slope may be used according to the variant of used elevation model of terrain relief (TIN or raster models). The result of the calculation is a layer of inclination of slopes. In the raster layer, the inclination of slope is given as a value of the given pixel, in TIN layer, this inclination is given as characteristics of an appropriate triangle. In order to simplify the further procedure, only the raster model of inclination of slopes will be considered. For the calculation of distance, it is possible to use the tool for calculation of Euclidean distance whose outcome is a raster field. The value of each pixel in such field is Euclidean distance of the center pixel from the nearest source, for instance communication, water stream, etc. The following pictures show examples of raster layer of inclination of slopes (Fig. 7.8), distance to communications (Fig. 7.9) and distance to waters (Fig. 7.10).

For the analysis of suitability of soil, the code of soil type from the soil database (Novák, 2000) was used which gives classification of soil types according to Table 7.3.

By the conversion function the flat objects of compounds of soil types are transformed into a raster file in which the code of soil type is the value of the pixel. The

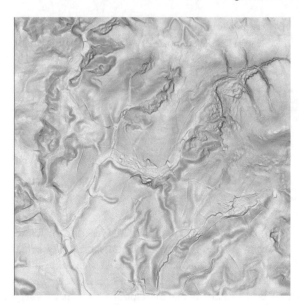

Fig. 7.8 Raster layer of inclination of slopes analyzed from the raster digital elevation model (DMR4)

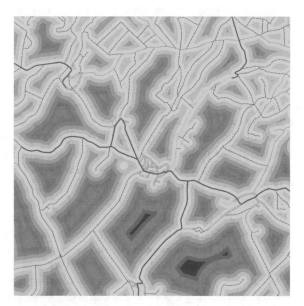

Fig. 7.9 Raster layer of Euclidean distance to communications

result of such transformation can be seen in the picture (Fig. 7.11). Similarly, it is possible to get a raster file from the vector layer of vegetation in which the value of the pixel is a code of vegetation type given in Table 7.4.

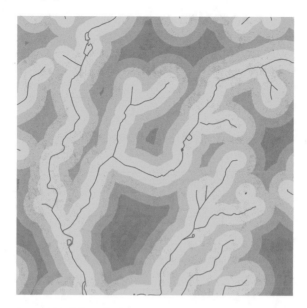

Fig. 7.10 Raster layer of Euclidean distance to waters

Table 7.3 Codes of soil types according to Novák (2000)

Code	Soil type
001	Weak soil (sandy)
002	Lighter moderate soil (loam-sandy)
003	Moderate soil (loam)
004	Stronger moderate soil (clay-loam)
005	Strong soil (clay)
996	Built-up area
997	Water area
998	Not defined

From the preceding text it is obvious that values of pixels in all entry layers are given by the method of their transformation into a raster file. That is why it is necessary to set how they will be mutually evaluated from the point of their influence on the complete analysis. One of the ways how to do it is to set number values into classes in each map layer, i.e. to reclassify them (Fig. 7.12).

The values of pixels of each layer will be divided depending on the level of suitability for the given purpose. The division is done in a way that the original values are put into e.g. ten classes where numbers 1–10 become new values of the individual pixels. Number 10 represents the best value and number 1 on the contrary its worst value. This way is called *suitability scaling*. Scaling may be done automatically or manually.

Fig. 7.11 Raster layer of soil types

Table 7.4 Codes of vegetation types (MoD-GeoS, 2010)

Code	**Vegetation type**
EA030	Forest nursery
EA040	Orchard
EA050	Vineyard
EB010	Pasture, meadow
EB020	Scrubs
EC015	Forest
NoData	No vegetation

Automated scaling is usually used with continuously changing values, in this case for Euclidean distances or inclination of slopes, while for setting the interval limits it is possible to use methods of natural breaks, method of quantile, fixed intervals, or it is possible to set the interval limits manually. The choice of a method depends on the division of frequency of values. The picture (Fig. 7.13) shows an example of setting the classification classes by the method of fixed limits.

The automated reclassification is then realized by a tool for reclassification (Fig. 7.14). In case of the distance from communications, the most suitable position of the modelled logistics center is as close to any communication as possible. On the contrary, if the automated reclassification is done for the distance to waters, then the most suitable position lies in the maximum distance from a water source.

Fig. 7.12   Raster layer of vegetation types

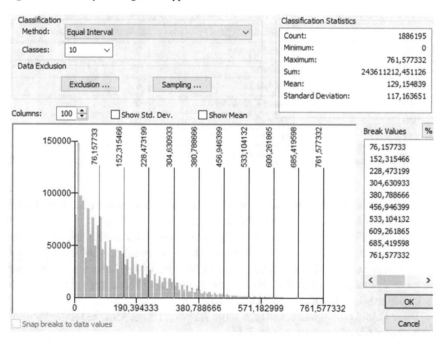

Fig. 7.13   Division of frequency of pixel values in a raster field of Euclidean distance from communications and automated classification into 10 classification classes

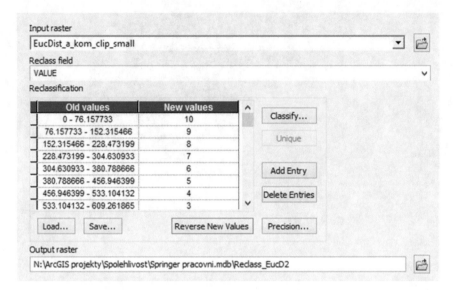

Fig. 7.14 Table of reclassification of the automated procedure for the distance to communications

Table 7.5 Codes of vegetation types and their suitability for deployment of logistics center

Code	Vegetation type	Suitability coefficient
EA030	Forest nursery	1
EA040	Orchard	1
EA050	Vineyard	1
EB010	Pasture, meadow	7
EB020	Scrubs	6
EC015	Forest	1
NoData	No vegetation	10

Identification NoData, which represents a place without any value, is attributed to cells that will not be included in the calculations.

Manual scaling is chosen in a case when enumeration (nominal) values are used in the classification. In the model example from the point of view of vegetation type, there are suitable places without vegetation, acceptable places are pastures and meadows, possibly also scrubs. Forest covers including forest nurseries, orchards and vineyards are considered to be unsuitable.

These values are implemented into Table 7.5 which is used for the reclassification of the raster file of vegetation (Fig. 7.15).

Similarly, also reclassification according to suitability of the raster file of soil types is done with the use of Table 7.6.

The results of scaling of the individual entry layers to the unified evaluation scale are obvious from the pictures (Figs. 7.16, 7.17, 7.18, 7.19 and 7.20).

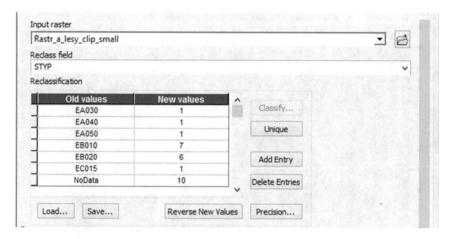

Fig. 7.15 Table of reclassification for manual procedure

Table 7.6 Codes of soil types and their suitability for deployment of logistics center

Code	Soil type	Suitability coefficient
001	Weak soil (sandy)	10
002	Lighter moderate soil (loam-sandy)	10
003	Moderate soil (loam)	9
004	Stronger moderate soil (clay-loam)	8
005	Strong soil (clay)	7
996	Built-up area	1
997	Water area	1
998	Not defined	1

Pixels of in such way prepared layers have the same influence on the result and they differ only internally by their value. However, as the individual evaluation criteria have various weights (Table 7.2), it is necessary to consider these weights also in the final multi-criterial analysis. The resulting file, so-called cost map, is created by the combination of partial reclassified files. For the calculations it is possible to use the tool of map algebra or weighted overlay (Fig. 7.21). The final result of the process model is shown in the picture (Fig. 7.22).

The cost map enables to evaluate the level of suitability of the pre-selected area for deployment of logistics center. The final decision about the deployment of the center is now in the hands of the responsible worker (for instance commander) who in such case may even undergo an acceptable risk, i.e. that the given space will not be optimal but that deployed units will be able to perform their activities even though they might be limited by this environment. As the complete procedure of analysis is usually created by formation of geo-process model, it is possible to deliberately

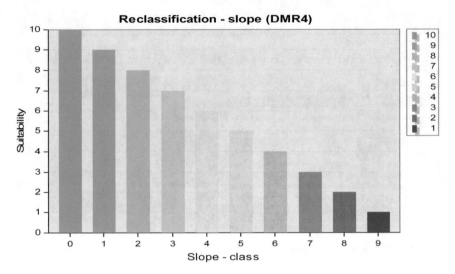

Fig. 7.16 Reclassification of inclination of slopes

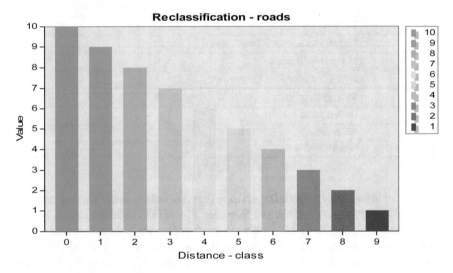

Fig. 7.17 Reclassification of distances to communications

modify the input conditions, obtain several results and decide which result will be used in the real conditions.

   The described procedure fully respects general principles of multi-criterial analysis including the creation of variants of solution. However, if there is a situation when it is not suitable or possible to decide according to the variants, an alternative way of multi-criterial analysis is chosen, which evaluates all factors only by comply—not comply. In such case, intermediate values are omitted and remain only places that

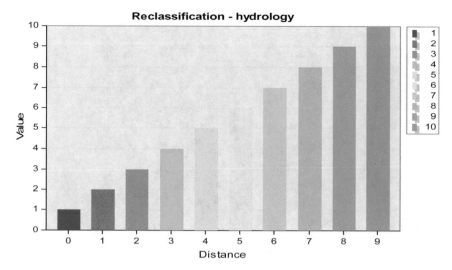

Fig. 7.18 Reclassification of distances to waters

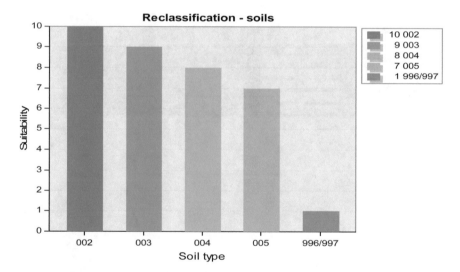

Fig. 7.19 Reclassification of soil types

comply or do not comply with the selection condition. The final result is obtained in the form of Boolean algebra when suitable places get attribute *Yes* or *No*. The resulting map of suitable and not suitable places obtained by this procedure is shown in the picture (Fig. 7.23). The same data as in the preceding case were used for its creation. The picture clearly shows that the resulting area of available places is smaller.

If it is possible, it is necessary to verify the obtained results, ideally by a personal reconnaissance. The last step in geo-process modelling is its implementation into

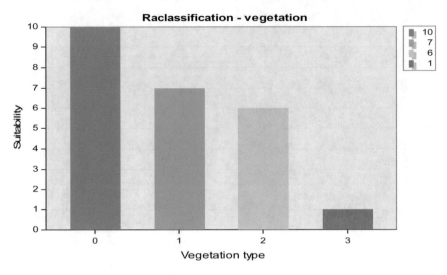

Fig. 7.20 Reclassification of vegetation

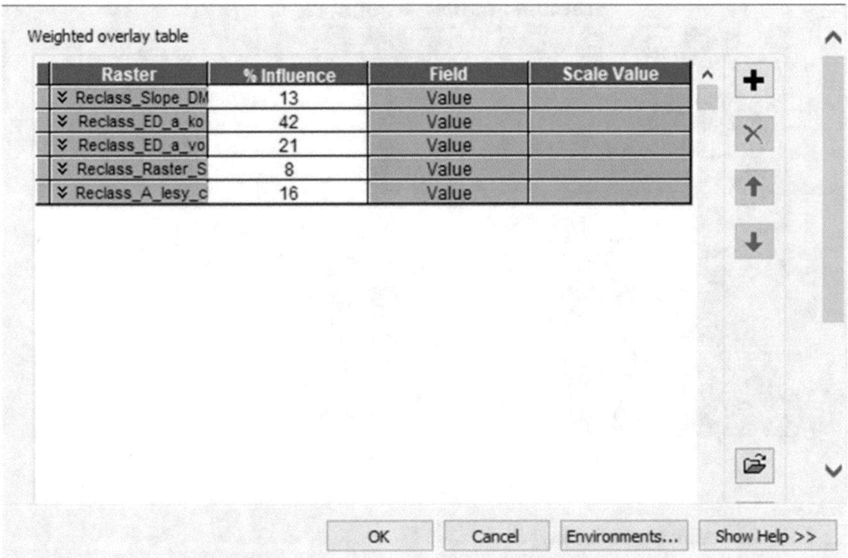

Fig. 7.21 Example of setting weights of entry layers with weighted overlay

practice. Such created model may be used for a one-time analysis or as a parametrized model for creation of analyses for the same or similar situation in various geographic conditions. In the described example, the results of the analysis are provided as a recommendation of a suitable space to the superior (e.g. within a briefing of the commander and their subordinates and cooperating units).

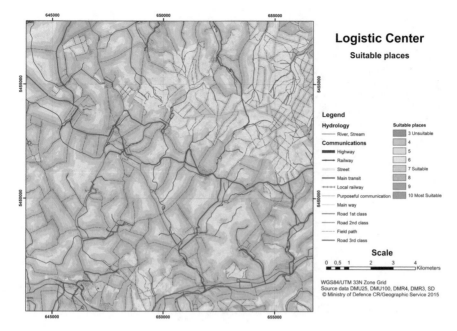

Fig. 7.22 Cost map for evaluation of suitable places for deployment of logistics center from data of DMU25, DMR4 and SD

7.2 Quality of Geographic Data in Multi-criterial Analysis

Results of a multi-criterial analysis of course depend on the quality of the used digital geographic data. Within the step of research of input data, it is necessary to evaluate not only the content of data files but also their utility value. The following text presents the influence of quality of background data on the resulting analysis whose geo-process model was presented in the previous paragraph.

In order to establish the influence of quality of the background data on the given analysis, the same analysis with less detailed background data with lower utility value was made. For the analysis positional database DMU100 was used, elevation model DMR3 and the same SD. Positional database DMU100 corresponds to the same standards as DMU25, its level of detail however, is on the level of a topographic map 1:100,000. The position of objects is then by one order lower. The elevation model DMR3 is again created by a network of height points, but unlike DMR4, this network is 10 × 10 m. That is why the results of the performed partial analyses correspond to the characteristics of accuracy. With regards to a thinner network of height points, the complete analysis was made with the size of a raster cell of 10 × 10 m. The pictures show results of the complete analysis with consideration of weights of the individual factors (Fig. 7.24) and with analysis of only suitable or unsuitable places (Fig. 7.25).

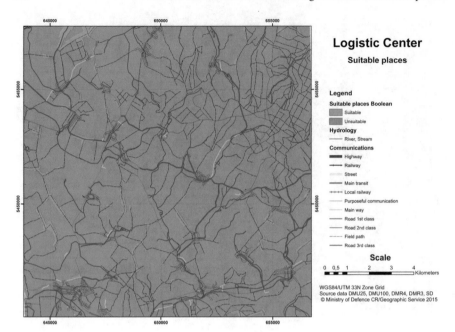

Fig. 7.23 Cost map for evaluation of suitable and unsuitable places for deployment of logistics center from data of DMU25, DMR4 and SD

A clear difference in obtained results can be understood from the pictures. This difference occurs when viewing a detail of area with the size 1 × 1 km. The legend in both pictures is the same as in the previous pictures (Figs. 7.26 and 7.27).

Differences in the values of particular pixels are evident from the picture (Fig. 7.28), where there are values obtained from less quality data (DMU100, DMR3) deducted from the values of individual pixels of analysis with the use of more accurate data (DMU25, DMR4).

From the example it is clear that the more quality data, the more quality analysis results. However, in a decision-making process the requirements for maximal accuracy of analysis are often second-rate, in some cases too accurate results may be unnecessary. For making the decision, the factor of time is often more important. It means the time from the instruction to start the analysis to the delivery of results. In this sense, it is necessary to consider not only the result itself, but also the required time for its obtaining.

With the increasing quality of the background data and increasing detail of partial analyses as well as the resulting analysis, also the requirements for time of calculation increase. In the given model example, the size of a raster cell for more accurate data 5 × 5 m was used, for less accurate data it was 10 × 10 m. In the same area, it is then necessary to work with four times bigger volume of data in the first case. This difference manifests in calculation times of program systems that are used for the analysis. Of course, for measuring times for calculations it is necessary to take into

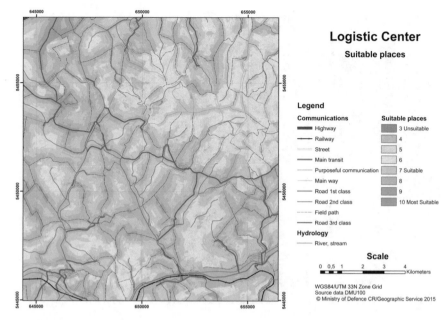

Fig. 7.24 Cost map for evaluation of suitable places for deployment of logistics center from data of DMU100, DMR3 and SD

account also the quality of used computing tools and the whole used system including consideration of simultaneously running programs (Schmid, Galicz, & Reinhardt, 2015). Yet it is recommended to consider these times, especially if the result must be provided within certain time.

Pokonieczny in his work (Pokonieczny & Mościcka, 2018) measured the calculation time needed for an analysis of movement possibilities of military units in an area of 60×60 km with the help of data VMap1 (Vector Map level 1 – similar to MGCP). Table 7.7 gives needed times for preparation of data and the calculation itself.

If an analyst has only limited time, he/she needs to provide the responsible worker—commander—with information about the quality of analysis result in the given time. Then it is the commander's decision whether a lower quality of results is sufficient for making the decision. If the given quality is not good enough, he/she needs to decide whether to shrink the area of analysis or to provide longer time to process the analysis.

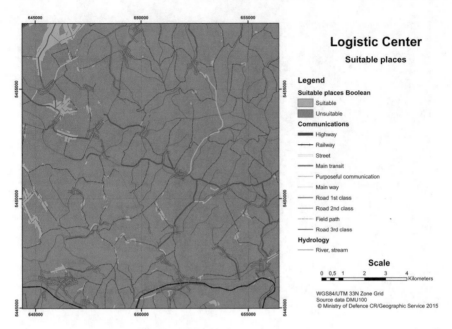

Fig. 7.25 Cost map for evaluation of suitable places for deployment of logistics center from data of DMU100, DMR3 and SD

Fig. 7.26 Cost map for evaluation of suitable places for deployment of logistics center from data of DMU25, DMR4 and SD—segment 1 × 1 km

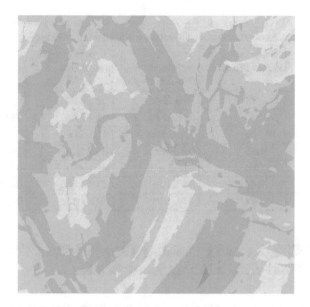

Fig. 7.27 Cost map for evaluation of suitable places for deployment of logistics center from data of DMU100, DMR3 and SD—segment 1 × 1 km

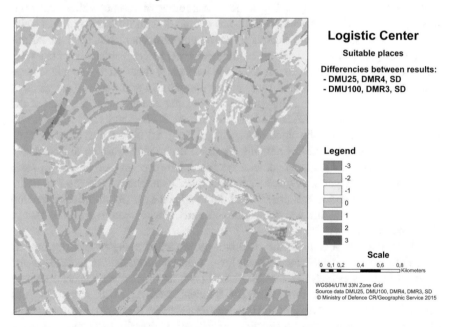

Fig. 7.28 Differences in values in Cost maps created from DMU25 and DMR4 and DMU100 and DMR3. The same SD was used

Table 7.7 Times for preparation and performance of an analysis of movement possibilities of military units in an area of 60×60 km [adapted from Pokonieczny and Mościcka (2018)]

Pixel size (m)	Number of pixels in area of interest	Time for preparation of data for analysis	Time of analysis calculation	Total time
100	295,772	90 h 22 min	2 h 28 min	92 h 50 min
200	74,317	22 h 42 min	0 h 37 min	23 h 20 min
500	12,060	3 h 41 min	0 h 6 min	3 h 47 min
1000	3085	0 h 57 min	0 h 2 min	1 h 0 min
2000	808	0 h 15 min	0 h 1 min	0 h 16 min
5000	147	0 h 3 min	<0 h 1 min	0 h 4 min
10,000	44	0 h 1 min	<0 h 1 min	0 h 2 min

7.3 Modelling Terrain Passability Using off-Road Vehicles with Multi-criteria Analysis Application

Modelling off-road vehicles movement in the field is one of specific tasks for using multi-criteria analysis in geoprocessing models. Possibilities of vehicle movement within a specified area are a key issue in a number of rescue, police and military operations. The intervention units are equipped with different types of vehicles (wheeled, tracked) with various power and traction characteristics and various equipment (tanker fire-engines, emergency vehicles, rescue vehicles, combat vehicles, etc.). The intervention area is relatively frequently found in the open landscape where the access from adapted roads is not possible. In addition, the activities of intervention units are affected by particular weather conditions at the time of the intervention. Intervention commanders have to decide which units and vehicles will be used in order to meet the objectives specified. Such a decision must often be taken under time pressure. Commanders mostly make decisions according to knowledge of local situation and limits of vehicles specified. However, in case of a larger-scale intervention in a larger area where units from other locations are present, it is advisable for the intervention manager to find out in advance the possibilities of moving particular vehicles within a specified area at particular meteorological situation in day or night time.

The vehicle movement in the field can be modelled using processing models with multi-criteria analysis methods application. Many specialists, organizations and research teams especially in the field of defence and security have been dealing with the issue for a long time; there can be mentioned, e.g., U.S. Army Corps of Engineers—Engineer Research and Development Center (ERDC) or the Department of Military Geography and Meteorology (DMGM), University of Defence in Brno, Czech Republic. Work effort of research teams resulted in The Next Generation NATO Reference Mobility Model (NRMM), which was presented for instance in Alvin and Haley (1992) and Jayakumar and Dasch (2016).

A team of specialists in the field of military geography and meteorology from the above mentioned department DMGM has long been involved in research focused on the vehicle movement in the field in terms of military vehicles (Rybansky & Vala, 2009; Rybansky, Hofmann, Hubacek, Kovarik, & Talhofer, 2015). The following text focuses on a brief description of the terrain passability model solution.

7.3.1 Initial Terms of the Solution for Vehicle Movement in the Field

The entire activity of units within a particular area is affected significantly by external circumstances, including the natural environment; those are both geographical conditions and meteorological conditions, or climatic conditions, resp. The effect of the environment can be assessed from many points of view in terms of the activity carried out.

When moving along roads, there are assessed their permanent properties such as quality and traffic performance of roads used, their configuration, their critical locations, e.g., bridges, bottlenecks or higher slope segments. However, there are also assessed road characteristics and their adjacent areas in terms of the task being performed. In this case, there are assessed temporary features such as barring or road blocking due to large-scale tree windfall after windstorms, road blocking with numbers of crashed vehicles in case of mass accidents, flooded roads at floods, etc.

When units move in the open field, several factors have to be evaluated; those are, particularly, slope gradients, number and shapes of micro-relief elements, surface types and their cover, vegetation species and distribution, depths and widths of water obstacles, surface condition resulting from current and former meteorological situation. Even in this case, all factors have to be assessed in view of the situation and the intended or performed activity.

The situation, in which the activity is accomplished, is a separate factor. In case the rescue units are deployed, in particular fire-fighters, those are areas struck by fire, landslides or mentioned mass traffic accidents. In case of military units deployment, it is particularly a tactical situation in the area specified by the activities both of the own troops and the activities of the enemy.

Possibilities how units may be used the field can be pre-modelled within GIS. Such models are designed for simulation and simulator technologies; those are particularly used for training commanders and staffs (Hubacek, 2011) as well as for the preparation and management of events.

Off-road behaviour patterns in natural environment have basically three levels. The first level is *a physical model* that characterizes the behaviour of a particular weapon, vehicle, aircraft, etc., in partial geographical and meteorological conditions, which are partially assessed. It is based on tactical-technical data and operational characteristics of the system. The highest level of physical models is a model of complex effect of natural environment on a given technique or system. Physical models are research results presented, e.g. in Rybansky (2009), Rybansky and Vala (2010) or are listed in normative acts (STANAG 2999, 2012).

In order to use physical models in geo-information technology environment, first, the *mathematical models* have to be created from them. Mathematical models consider not only principles in terms of vehicle movement but also applied geographical and meteorological data and their accuracy and reliability characteristics. Mathematical modelling used in command and control systems are dealt with by a number of authors both from the Czech Republic and abroad (De Felice & Petrillo, 2013; D'Amico, Di Martino, & Sessa, 2013; Hošková-Mayerová, Talhofer, & Hofmann, 2013). In mathematical modelling, it is inevitable to consider the uncertainty of the data themselves given by their natural characteristics or to work with their positional and attribute accuracy (Svatonova & Rybansky, 2014; Veregin, 1999; Talhofer, Hoskova, Hofmann, & Kratochvil, 2009). Data uncertainty is appropriate to include into models themselves, e.g., by using the general theory of fuzzy sets (Zadeh, 1965) since in many program systems for geo-information processing, there are appropriate tools available for data uncertainty solutions (Kainz, 2007; Di Martino & Sessa, 2011; Hofmann, Hoskova-Mayerova, & Talhofer, 2013).

Mathematical models are transformed into *information models* as a rule for particular computing and programming environment, e.g., ArcGIS by ESRI (2013). There it is possible to create complex process diagrams working on given geographical and meteorological data.

However, all the models have to be verified thoroughly, both by partial and complex field tests. Based on the test results, models at all levels have to be modified so that they match the reality as much as possible.

The following text offers a brief description of modelling techniques for movement of off-road vehicles in the field.

7.3.2   Physical Modelling of Terrain Passability

The process of physical modelling of a vehicle movement in the field is based both on domestic and foreign literature sources. In the Czech Republic, Rybanský and his co-workers have particularly been engaged in long-term work focused on analysing and verifying effects of single geographical elements on a vehicle movement. He elaborated the basic theory of this modelling (Talhofer, 2004) and on its base also the methodology or assessing characteristics of military technique, both wheeled (ArcGIS, 2015) and tracked (MoD-GeoS, 2010). A similar principle is used for procedures applied particularly in the US ground forces, which are being developed in ERDC.

The physical model consists in the driving characteristics of a particular type of a vehicle; those are specified by a number of technical parameters (tactical-technical data, TTD). The most basic ones are as follows:

Table 7.8 Selected tactical-technical data for military vehicles

Vehicle type	UAZ 469 (personal vehicle)	Tatra 815 8 × 8 (heavy-duty truck)	BVP 2 (armoured tracked vehicle)
Length [m]	4	8.95	6.73
Width [m]	1.8	2.50	3.15
Height [m]	2	3.95	2.45
Weight [kg]	2400	21,000	14,000
Maximum climbing ability [°]	30	30	35
Maximum speed on the road [kmph]	105	85	65

- Dimensions and weight of a vehicle,
- Engine performance parameters,
- Chassis construction type (tracks, wheels),
- Number and type of tyres; track width and the size of the effective area, resp,
- Climbing ability, cross-over ability, angles of approach, longitudinal and transverse roll and many others.

Although TTD are declared by a manufacturer or a technique supplier, resp., before putting the technique into operation they are verified and completed with operation tests in authorized institutes: rescue technique—The Technical Institute of Fire Protection (TIFP, 2015) or The Military Technical Institute of Ground Forces (VTUPV, 2010). An example of selected TTD for type vehicles is presented in Table 7.8.

Actual real driving characteristics of a particular vehicle are also given by the state of driving units, chassis groups, tyres or tracks. Driver's experience and his/her current physical and psychic condition are equally important factors.

Driving characteristics in the field will be reflected in the vehicle's ability to manoeuvre in a particular environment under specific external circumstances. This ability is closely related to the terrain configuration and characteristics of its individual components. The entire terrain components and phenomena form a specific character of a terrain or landscape by their action and mutual interference. Many elements are interlinked; therefore, their effect on movement of vehicles have to be considered as well in terms of a complex view of the terrain. In case that the vegetation effect on a vehicle mobility is considered, the consideration has also to be given to reliefs, soils and further components affecting the vegetation species composition. However, such a complex view is rather complicated; therefore, it is appropriate for the modelling purpose itself to divide the environmental effect on a vehicle movement into single components, and create the models themselves using the synthesis of these data from the resulting data.

As stated by Talhofer (2004), the landscape effects entering the model of a vehicle movement can be characterized using partial factors and those can be expressed as

Table 7.9 Major deceleration coefficients

Major coefficient	Geographical significance and effect
C_1	Terrain relief (slope gradient, micro-relief shapes)
C_2	Vegetation cover
C_3	Soils and ground cover
C_4	Meteorological and climate conditions
C_5	Hydrology
C_6	Built-up areas
C_7	Road networks

deceleration coefficients of a particular type of a vehicle in relation to its maximum speed on the consolidated road (Table 7.9).

All major coefficients are categorized into partial coefficients describing in detail partial effects of the natural environment (Table 7.10).

The individual coefficients C_1–C_7 are calculated by using an appropriate function from partial coefficients in each group; the resulting coefficient is a function of those coefficients, and take values from 0 to 1. The resulting model speed of a vehicle movement is, besides a maximum speed of a vehicle, the function of all mentioned coefficients:

$$v = f(v_{max}, C_1, C_2, \ldots C_7) \tag{7.1}$$

In order to analyse the effect of all geographic factors on the technique movement, first, their detailed classification have to be carried out, it has to be determined, which criteria are used for coefficients assessment, and it is necessary to specify the initial values of deceleration coefficients due to the effect of individual factors.

In order to verify this classification, extensive field tests had been accomplished to measure the soil load capacity, vehicle ability to overcome various barriers and vegetation obstacles, in particular trees and dense shrubs. The main results from experiments are measured and computed coefficients values for off-road vehicles deceleration at overcoming various relief and barrier types (Talhofer, 2004), methodology for assessing the soil load using digital soil maps (Hubacek, Almasiova, Brenova, Bures, & Mertova, 2016), and determined dependencies of a vehicle movement speed on various types of both consolidated and unconsolidated surfaces (Hoskova-Mayerova, Hubacek, Bekesiene, & Bures, 2017). Resulting from presented coefficients, there are determine criteria for assessing possibilities of movement for a particular vehicle in the field, i.e., under what conditions the partial factor does not restrict the movement at all, when the movement is restricted, and when the factor prevents the movement. However, due to the complexity of relationships among individual factors, many coefficients are constantly refined because of further additional measurements and tests. This applies particularly to the soil, vegetation and meteorological conditions.

Table 7.10   Categorization of major deceleration coefficients

Major coefficient	1st order coefficient	Coefficient significance
C_1	Terrain relief effect	
	C_{11}	Slope gradient effect
	C_{12}	Micro-relief effect
C_2	Vegetation—woody species effect	
	C_{21}	Trunk distance effect
	C_{22}	Trunk diameter effect
C_3	Soil and ground cover effect	
	C_{31}	Soil effect
	C_{32}	Vegetation cover effect
	C_{33}	Surface roughness effect
C_4	Weather and climate effect	
C_5	Water courses and areas effect	
	C_{51}	Effect of a water course or water area
	C_{52}	Effect of a water course or water area depth
	C_{53}	Effect of a water course width
	C_{54}	Effect of a water course flow rate
	C_{55}	Effect of bottom characteristics
	C_{56}	Effect of banks characteristics
C_6	Built-up area effect	
	C_{61}	Effect of block built-up area
	C_{62}	Effect of villa built-up area
	C_{63}	Effect of cottage built-up area
	C_{64}	Effect of gardens and gardening colonies
C_7	Roads effect	

Illustrative examples of testing partial factors on movement of military vehicles in the field are presented in Figs. 7.29, 7.30, 7.31, 7.32, 7.33 and 7.34.

7.3.3   Mathematical Modelling

The mathematical model is based on the above-characterized physical model and quality parameters of source data used. Inasmuch as the vehicle movement models have to be solved using the data of various quality, the model for calculating individual deceleration coefficients has been proposed: it considers the data uncertainty. The model is based on the theory of fuzzy sets because those make possible to optimize

Fig. 7.29 Light off-road vehicle during test of tractive power (*Photo* Rybansky 2017)

Fig. 7.30 Test of achieving of critical load capacity (*Photo* Talhofer 2017)

Fig. 7.31 Tatra 815 8 × 8 truck in the ability to overcome transverse waves (*Photo* Zerzan 2015)

calculations with uncertain elements both in position and in their characteristics (Zadeh, 1965; Kainz, 2007).

Fig. 7.32 Testing of tree penetration and movement in the forest by BVP 2 (*Photo* Rybansky 2017)

Fig. 7.33 Measurement of load soil capacity by penetrometer E-960 (*Photo* Zerzan 2017)

An example of using fuzzy sets is shown in the Figs. 7.34 and 7.35. In order to determine the deceleration coefficient value due to the surface, there are considered both the roads and the open field. The applied database DMU25 comprises the information on the road width only for highways and roads, forest field path is recorded using their axis regardless the fact their real width is usually 4 m. In case that further

Fig. 7.34 Testing critical values when overtaking the micro-relief shape of Tatra 815 8 × 8 truck (*Photo* Dohnal 2017)

Fig. 7.35 Illustration of a raster route record format in a superposition with an orthogonalized aerial photo. Red points are records of a Tatra 815 8 × 8 truck position at field tests

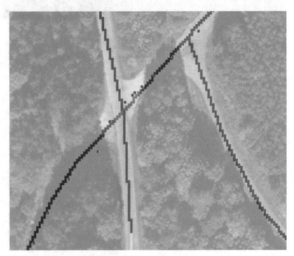

calculations require transformation of vector records of roads into raster format (in this case, 1 m as pixel size), the road is modelled as a sequence of pixels carrying the information "Path"; outside them, in other pixels of a raster field there is the information "No path". This way, so called sharp sets are obtained. However, picture (Fig. 7.35) shows that this approach do not corresponds to the reality because the actual position of a vehicle is frequently off the road model regardless the fact it moved along the road.

Due to the fact that it is not possible to determine accurately the real position of a road using the data stored in a database, its probable position is to the right and to the left of this axis within 5 m distance. The limit of 5 m right and left of the road was determined because of the road assignment to a given object geometry accuracy group specified by the ACC code.

In DMU25, the object geometry that records its position, is given in two accuracy classes: accurate and approximate, which corresponds to the declared accuracy for a given object class expressed by a mean position error. In case of road network object, this mean position error is 5 m (accurate position) and 15 m (approximate position).

Table 7.11 Table of object attributes and their values—part of CTO issued by VGHMUr (2015)

Code	Name	List of values	Notes
ACC	Quality of horizontal delimitation	001 accurate	Follows the product criterion
		002 approximate	Exceeds the product tolerance
		003 uncertain	Position estimated
		007 rigorous	Position more accurate in order than the product requirement

Fig. 7.36 Superposition of a recalculated cost map using fuzzy sets (grey scale) of a recorded Tatra 815 8 × 8 truck position and orthogonalized aerial photo

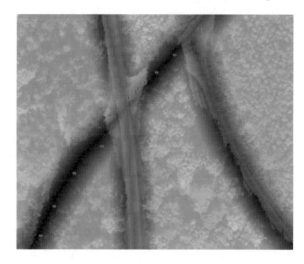

For each object group, this general declaration is refined by the ACC code, which informs about the real quality of the horizontal delimitation (Table 7.11).

The ACC code for the data used in the example was 001, i.e., the mean position value error for field forest paths is 5 m. Therefore, the fuzzy set theory can be used to solve the path position identification problem. The pixel-to-path value will be the maximum in the path axis and will drop to the right and left of it to a distance of 5 m. The result of the path position transformation using a fuzzy-method is shown in Fig. 7.36.

When calculating the individual deceleration coefficients, the three basic limits are usually considered as movement possibility criteria; considering this parameter, the terrain is completely passable (deceleration coefficient is 0), there is restricted passability (deceleration coefficient is bigger than 0 and smaller than 1) or it is completely impassable (deceleration coefficient is 1). Passability classification in the restricted passability terrain is proportional to the scattering of values of a particular parameter within declared positional and attribute accuracy. The further text presents basic methodological approaches to calculating individual coefficients (Hofmann, Hoskova-Mayerova, & Talhofer, 2013), which had been used in a pilot model version. However, the particular model was modified resulting from real parameters of a specific technique; this concerns particularly the specification of limit deceleration values for a specific vehicle.

The coefficient of *terrain relief effect* C_1 is very complex; it comprises both the slope gradient effect and micro-relief effect (ditches, embankments, terrain grades, etc.). The coefficient is calculated by the initial relationship:

$$C_1 = \prod_{j=1}^{2} C_{1j}. \tag{7.2}$$

The coefficient of slope gradient effect C_{11} can be expressed using the equation:

$$C_{11} = \begin{cases} \frac{gradT_{max} - SH}{gradK_{max}}, & \text{for } SH \leq gradT_{max} \\ 0, & \text{for } SH > gradT_{max} \end{cases}, \tag{7.3}$$

where

- $gradT_{max}$ is the maximum angle of gradient in the open terrain,
- $gradK_{max}$ is the maximum angle of gradient on the road,
- SH is the mean angle of gradient.

All values are given in degrees with accuracy to the whole grade.

If the micro-relief affect is considered as a parameter allowing overcoming height or depth obstacles, then the coefficient C_{12} can be expressed as:

$$C_{12} = \begin{cases} 0, & \text{for } h > h_{max} \\ 0, & \text{for } w > w_{max} \\ 1, & \text{for } h \leq h_{max} \wedge w \leq w_{max} \end{cases}, \tag{7.4}$$

where

- h is height of a height obstacle (terrain grade) with accuracy to the tenths of metre,
- h_{max} is the height of the maximum surmountable height obstacle with accuracy to the tenths of metre,
- w is the width of a depth obstacle with accuracy to the tenths of metre,
- w_{max} je maximum surmountable width of a depth obstacle with accuracy to the tenths of metre.

All calculations are based on the TTD values of a particular assessed vehicle.

When assessing the *vegetation effect* using the coefficient C_2, vegetation is specified only as wooded areas, especially forest stand complexes or orchards, resp. The vegetation effect is calculated by the Formula (7.5):

$$C_2 = \prod_{j=1}^{2} C_{2j}. \tag{7.5}$$

The values of individual coefficients are expressed using relationships (7.6), which had been got by direct measurement in the field and supplemented with expert estimate:

$$C_{21} = \begin{cases} 1, & \text{for } sp \geq sp_{opt} \\ 0.5, & \text{for } sp = \left(sp_{min}; sp_{opt}\right), \\ 0, & \text{for } sp \leq sp_{min} \end{cases} \qquad (7.6)$$

where

- sp is a trunk distance expressed with accuracy to the units of metres,
- sp_{min} is a minimum passage profile for a particular vehicle,
- sp_{opt} is the optimal passage profile for a particular vehicle.

The coefficient C_{22} is calculated from a prevailing woody species trunk diameter at 1.2 m height above the ground. A trunk diameter is marked td and is expressed in thousandths of a meter with accuracy to units of centimetres; td_{opt} is a trunk diameter that is no obstacle for a particular vehicle and td_{max} is a maximum trunk diameter, which the vehicle is able to overcome. Coefficient values for particular vehicles were again determined at field tests and completed with expert estimates. The coefficient calculation can be expressed as a relationship

$$C_{22} = \begin{cases} 1, & \text{for } td \leq td_{opt} \\ 0.5, & \text{for } td = \left(td_{opt}; td_{max}\right). \\ 0, & \text{for } td \geq td_{max} \end{cases} \qquad (7.7)$$

The effect of soils and ground cover can be expressed using a simplified procedure applying basic soil categorization with dominating soil types in the Czech Republic. A simplified procedure results in an expert estimate of soil effect on vehicle deceleration, which had been based on the above-mentioned physical modelling and field tests.

The effect of soils and ground cover is expressed by the relationship:

$$C_3 = \prod_{j=1}^{3} C_{3j}. \qquad (7.8)$$

Coefficient C_{31} was determined according to Novák (2000) and Rybansky (2009) for the model of terrain passability and supplemented by an above mentioned expert estimate. Its values are in Table 7.12.

Table 7.12 Determination of the coefficient C_{31} for soil types and meteorological conditions (expert estimate)

Code	Soil type	Coefficient C_{31} for period		
		Dry	Moderate	Wet
001	Light soil (sandy)	0.5	0.7	1.0
002	Lighter soil (loamy-sandy)	0.9	0.7	0.5
003	Medium soil (loamy)	1.0	0.9	0.3
004	Heavier medium soil (clay-loamy)	1.0	0.5	0.1
005	Heavy soil	1	0.3	0

For the purpose of a passability model it is possible to define the period related predominantly to the soil moisture content and apply the following definitions (NDMC, 2017; CMeS, 2017; Novák, 2000):

– Dry period (agronomical drought) that is characterized with insufficient water content in the soil; it shows the low soil moisture caused by meteorological drought. Further remarkable effects are shown by soil characteristics and management practices,
– Moderate period is a period characterized with a total liquid precipitation up to 40 mm in three days in October–April period, and up to 60 mm in May–September period,
– Wet period is characterized with a total liquid precipitation over 40 mm in three days in October–April period, and over 60 mm in May–September,
– Coefficients C_{32} and C_{33} were not considered in the described phase of the model and their value was 1 for all cases.

Weather and climate effect expressed with coefficient C_4 has been simplified in the given task. Only the amount of liquid precipitation was considered in the given period of the last three days and was included in the calculation of the coefficient C_{31}. Total precipitation values were taken from meteorological stations measurements within the area evaluated.

Effect of water courses and water areas as water obstacles was evaluated in terms of terrain reliefs, especially considering micro-reliefs, including the banks characteristics expressed with the coefficient C_{56}. This paragraph presents the methodology for assessing the impact of water courses and water areas when characteristics of presented objects are considered as separate elements.

The aggregate impact of water courses and water areas is expressed as

$$C_5 = \prod_{j=1}^{6} C_{5j}. \qquad (7.9)$$

The individual partial coefficients are derived based on TTD and instantaneous conditions of a given water course or water area.

Table 7.13 Deceleration coefficient values based on the water obstacle type

C_{51} coefficient value	Type of water obstacle
0.0	Swamp, bog, silt, rapids, waterfall
1.0	Water course, canal, water ditch
1.0	Pond, water reservoir

Table 7.14 Results of wading measurements in a water course for Tatra 815 8 × 8 truck (Rybansky, 2009)

Parameter	Value		
Maximum fordability (m) at water course flow rate	≤ 1 m·s^{-1}	$1–2$ m·s^{-1}	≥ 2 m·s^{-1}
	0.9	0.8	0.7

The effect of a water obstacle type expressed with the coefficient C_{51} is shown in Table 7.13.

The coefficient C_{52}, specified as the impact of the water obstacle depth is considered in terms of wading. If the water depth is higher than the wading limit based on TTD, the coefficient value is 0; otherwise, its value is 1. However, it has to be considered that the wading value is related to the depth of a water reservoir. If a case of wading a water course occurs, the water course flow rate has to be considered as well.

The coefficient of a water obstacle width C_{53} is always 1 in terms of a solved task; unless the movement in water is affected by any other factor, for instance its depth, the width of a water obstacle plays no significant role there.

The possibility to overcome a water obstacle is also affected by a water course flow rate. The coefficient C_{54} values shown in Table 7.14 were defined based on TTD of Tatra 815 8 × 8 truck and accomplished experiments.

The coefficient of bottom characteristics impact C_{55} was always 1 in terms of a solved task in the first variant.

The coefficient of banks characteristics C_{56} is substantially the same as the coefficient C_{12} given in Formula (7.4). If the banks parameters are known (object status, its height, type and construction material), this formula can be used for its determination. If these parameters are unknown, then for water course wider than the vehicle cross-over ability based on TTD is, the coefficient C_{56} value equals 0; otherwise it will equal 1.

The effect of a built-up area was solved in the described variant in a simplified way. Assuming that the movement of vehicles within a built-up area is considered mainly on roads or off them—not in the open unbuilt area—the type of built-up area is not specified and all structures regardless their purpose are classified impassable. Then there will be:

– $C_6 = 0$ for all structures,
– $C_6 = 1$ for all areas excluding structures.

Table 7.15 Coefficients for road impact

Coefficient value C_7	Road types
0.87	Highways and expressways
0.83	Main road, main transit
0.73	Secondary roads
0.53	Other roads, streets, purpose roads, street tracks
0.44	Consolidated roads, field forest paths
0.00	Railroads, siding

The following simplification was used for calculating *the impact of roads* expressed by the coefficient C_7:

– Roads are valued by only one coefficient whose value is derived from the type of a road,
– Road network density is not considered,
– Except for fords, all technical structures (bridges, passes, etc.) have the same parameters as roads, which they belong to,
– Railroads are considered for an assessed vehicle as impassable,
– Road passability is assessed based on the methodology for water case impacts.

The values for the coefficient C_7 are shown in Table 7.15.

7.3.4   Information Model

For the specific solution, data from the Geographic service of the Army of the Czech Republic were used; those were topographic, elevation/height and thematic data in various forms and data formats. The information model was designed to allow assessing the impact of various quality height models on the resulting passability analysis, and at the same allows eliminating the influence of various quality position and thematic data on the entire passability model.

7.3.4.1   Data Used

Due to the established conditions for creating a model, one position database (DMU25), one thematic database (SD–Soil Database) and four height models (DMR3, DMR4, DMR5 and SRTM) were used.

DMU25 was a basic position database used for the solution; its description is presented in Sect. 4.2.1. This database was used to find out objects characteristics on

Table 7.16 Basic characteristics of applied height models [*sources* JPL (2015), VGHMÚř (2015), CUZK (2018)]

Height model	Data structure	Height mean error		
		Open terrain (m)	Structures	Wooded terrain (m)
SRTM	Regular point network $1 \times 1''$	16	16	16
DMR3	Regular point network 10×10 m	1–2	1–2	3–7
DMR4	Regular network 5×5 m	0.3	Not specified	1
DMR5	TIN	0.18	Not specified	0.3

the Earth's surface, which may affect the terrain passability. From the stored data, values of most C_{ij} coefficients presented in Table 7.10 are determined or calculated.

To express the soil type effect (coefficient C_{31}), the already mentioned soil database (Novák, 2000) have been used. The soil type surface areas have been taken from this database.

The terrain slope is of crucial importance for the terrain passability solution; it can be calculated using various digital height models. In order to assess the accuracy impact and resolution levels of various height models, four above-mentioned models available for the Czech Republic territory were used for calculations. Their summary characteristics are shown in Table 7.16.

7.3.4.2 Requirements for Data Used

In order for the model to work with the data that had entered the calculation, first, it was necessary to find out, whether the input data corresponded to the quality criteria in terms of the calculation considering both requirements for their completeness and technical quality. Table 7.17 summarizes basic requirements for input data quality in terms of the terrain passability model.

The table presents optimal quality requirements for data used. However, the model result is affected by the actually used data as it had been specified in model analysis example in Sect. 7.2.

Table 7.17 Database requirements

Coefficient	Coefficient significance	Data quality requirements (accuracy, availability, etc.)
C_{11}	Slope gradient effect	Mean height error value existence
C_{12}	Micro-relief effect	Mean error both in position and 0.1 m height
C_{21}	Trunk distance effect	Mean error in a trunk position of 1 m
C_{22}	Trunk diameter effect	Mean error in specifying a trunk diameter of 0.02 m
C_{31}	Soil effect	Soil type code existence
C_{32}	Vegetation cover effect	Vegetation type code existence
C_{33}	Surface roughness effect	Roughness type code existence
C_{51}	Effect of a water course or water area	Water course or water area type code existence
C_{52}	Effect of a water course or water area depth	Mean error at a depth of 0.1 m
C_{53}	Effect of a water course width	Mean error in a width of 1 m
C_{54}	Effect of a water course flow rate	Mean error at a velocity of 0.1 m·s^{-1}
C_{55}	Effect of bottom characteristics	Bottom characteristics code existence
C_{56}	Effect of banks characteristics	Existence of codes specifying status, height, type and material composition of a bank
C_{61}	Effect of block built-up area	Block built-up area code existence
C_{62}	Effect of villa built-up area	Villa built-up area code existence
C_{63}	Effect of cottage built-up area	Cottage built-up area code existence
C_{64}	Effect of gardens and gardening colonies	Garden and gardening colonies code existence
C_7	Effect of roads	Road type code existence

7.3.5   Process Modelling

Based on the mathematical model, a process analysis model covering the possibilities of moving vehicles in the field has been developed. The basic process scheme in shown in Fig. 7.37.

The process model was processed with a program system ArcGIS (Esri, 2013) using a vehicle, whose technical parameters are taken from the database of its TTD. The model also allows using data input of various quality.

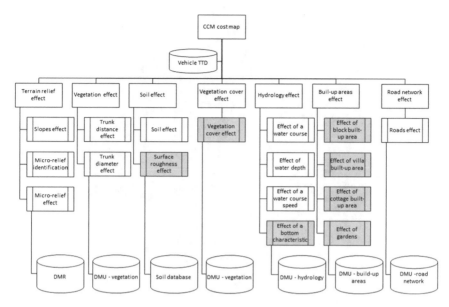

Fig. 7.37 Scheme of a process model for terrain passability solution. Brown coloured blocks are dealt with in a simplified way

The process model output results in a cost map in the form of a raster file. The size of a raster field corresponds to the resolution level of a height model used. The raster cell content is the model speed of a particular vehicle in a specific terrain area. Partial cost maps are first created for individual deceleration coefficients given by partial factors effects. Figure 7.38 is a schematic illustration of partial cost maps for major deceleration coefficients calculated for Tatra 815 8 × 8 truck from the Military training area Libavá.

From the individual partial cost maps, the total cost map for a particular vehicle is calculated using an appropriate Overlay function variant. Figure 7.39 shows an illustration of a cost map cut-out for a given vehicle from the same area.

In particular, the cost map allows judging whether a specific area is for a particular technique impassable, passable with difficulties or impassable. However, in addition to a cost map, there can be solved various tasks such as the calculation of an optimal route in the open field, optimal distribution of means for blocking passages, etc.

It is evident that cost maps calculated from different source data have different qualitative characteristics. Figures 7.40, 7.41, 7.42 and 7.43 present cost map cut-outs for Tatra 815 8 × 8 truck from the Military training area Libavá elaborated using a process model for the input data combinations presented in Table 7.18.

However, a cost map is only a statistical basis for assessing whether a particular vehicle has abilities to move within a given terrain area in terms of its technical parameters. Nevertheless, the actual movement of a particular vehicle within a specific area is also significantly affected by the driver's ability to control a vehicle as

Fig. 7.38 Example of partial cost maps for individual major deceleration coefficients. The colour scale is selected from green (passable terrain) to red (impassable terrain)

Fig. 7.39 Cost map cut-out for Tatra 815 8 × 8 truck

Fig. 7.40 Cut of cost map for Tatra 815 8 × 8, data combination code KSRTM

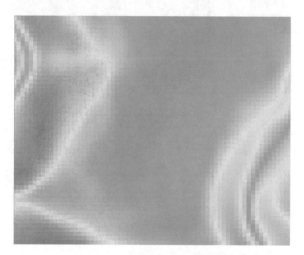

Fig. 7.41 Cut of cost map for Tatra 815 8 × 8, data combination code K3

well as by driving a vehicle in previous sections. However, it is necessary that the actual cost map or vehicle movement over the terrain could as accurately as possible express the vehicle's behaviour in real conditions.

Fig. 7.42 Cut of cost map
for Tatra 815 8 × 8, data
combination code K4

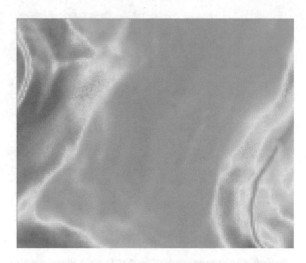

Fig. 7.43 Cut of cost map
for Tatra 815 8 × 8, data
combination code K5

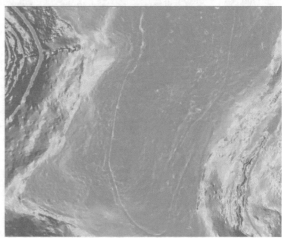

Table 7.18 Combinations of input data used

Data combination code	Input data used
KSRTM[a]	DMU25 + SRTM
K3	DMU25 + DMR3 + SD
K4	DMU25 + DMR4 + SD
K5	DMU25 + DMR5 + SD

[a]This combination has only been used for assessing the terrain relief effect

7.4   Process Model Verification

In order to assess the created complex models of vehicle movement, they have been subjected to three complex verification tests so far. The purpose of models verification was to verify the behaviour of the entire model in the environment, in which the units of the Army of the Czech Republic are located or can be located. The testing took place in a Military training areas (MTA) Březina and Libavá (2015, 2017). Test areas are presented in Figs. 7.44 and 7.45.

The main objective of the tests was to find out to what extent the results of modelling are usable in real application in decision-making processes in command and control systems; modelling was accomplished through above-mentioned physical, mathematical and information models.

In the preparatory parts of the tests, cost maps had been calculated using given areas for each tested vehicle available for a team. Within the tests, the following technique was verified:

– Personal off-road vehicles UAZ 469 (UAZ) and Land Rover 110 (LR),
– Medium truck Tatra 810 6 × 6 (T810),
– Heavy truck Tatra 815 8 × 8 (T815),
– Wheeled armoured carrier PANDUR II (PII),
– Light armoured vehicle IVECO (IV),
– Infantry armoured vehicle BVP 2 (BVP),
– Medium tank T-72M4 CZ (T72).

Field tests were accomplished 6th and 7th May 2014 (MTA Březina) and 4th–7th May 2015 and 10th–14th April 2017 (MTA Libavá). During the test, each vehicle passed through specified points, resp. in the specified direction several times at different times of the day and with general adherence to the set route. The actual routes travelled were recorded by the GPS receiver in a spot mode with a time interval of 1 s. Three Trimble GPS receivers with an external antenna Geoexplorer XT, XT3000 and XT6000 equipped with TerraSync software were used for recording.

The measured data were later corrected in a post-processing using a network of permanent CZEPOS reference stations and PathFinder software.

In total, over 2000 km routes have been recorded mostly in open terrain with minimum consolidated roads. From the corrected GPS data, discrete points of eth routes travelled were obtained with the instantaneous values of their coordinates in WGS84/UTM, the UTC time, the distance travelled, horizontal speed and the speed on the physical surface at each point. These points were transformed into raster format with a pixel size of 1 × 1 m with pixel horizontal rate.

For all routes recorded, the differences between the model and real speed of a specific vehicle were found in a given combination of data in a given cost map (Fig. 7.46).

To display a spatial distribution of velocity deviations, it is possible to use a deviation chart with the terrain relief profile displayed. As an example, the route of Tatra 815 8 × 8 truck was selected; the vehicle moved both on the consolidated road

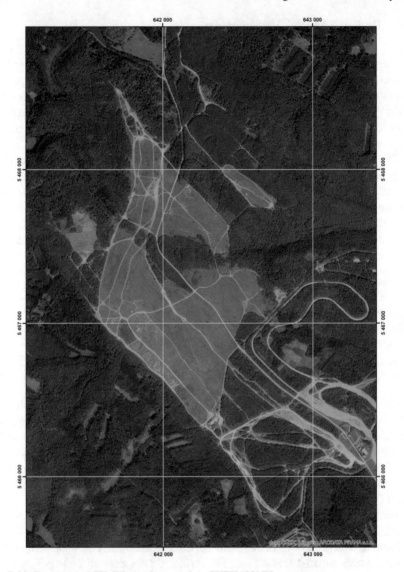

Fig. 7.44 Test area in MTA Březina [*source data* CUZK (2017)]

an in the open field covered with sparse shrubs; the route length was 5.7 km. An example of the route is in Fig. 7.47.

Figure 7.48 lists the speed deviations for the data combination K4—DMU25, DMR4 and ÚDP.

To understand the causes of differences, it is necessary to examine their statistical distribution; it is shown in Fig. 7.49. Axis *x* shows differences between modelled and actual spead. The *y* axis shows the number of cases.

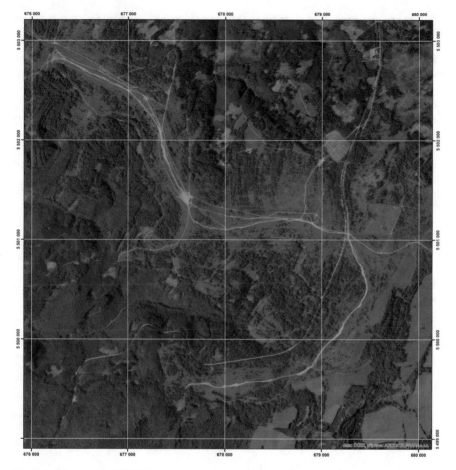

Fig. 7.45 Test area in MTA Libavá [*source data* CUZK (2017)]

Similar distributions of speed deviations are also apparent when using the remaining data combinations. The distribution of deviations shows two distinct local extremes. To identify the causes of local extremes, an analysis of the areas where deviations reach typical values was performed. Figure 7.50 identifies the sections where the speed differences belonged to the first or the second significant local extreme resp. These sections, in fact, corresponded to areas where the vehicle crossed the open terrain (the first local extreme) and when travelled along the roads, in this case, unconsolidated ones (the second local extreme).

The distribution of deviations between modelled and actual speed in analyzed sections in the data combination K4 is evident from the following charts (Figs. 7.51 and 7.52). In both cases axis x shows differences between modelled and actual spead. The y axis shows the number of cases.

Fig. 7.46 An example of a difference between the calculated speed in the model (red figures) and the actual speed (black figures)

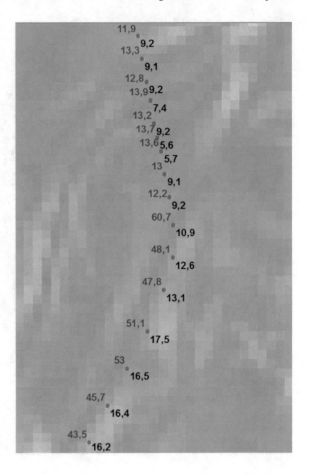

A similar distribution is also for the remaining data combinations. The distribution showed relatively high degree of coincidence of deceleration coefficients in the open field; however, the coefficients had to be refined for consolidated roads.

7.5   Model Simplification

A described model of the terrain passability addressed the effect of most geographical factors on the movement of vehicles in terms of multi-factorial analysis. However, its field tests verification showed that due to its complexity, it is difficult to find out which factors are independent of each other, and which ones, on the other hand, are functionally dependent on each other. In order to find functional dependences among factors, first, it is necessary to simplify this model. It is advisable to divide the complex model into sub-parts and verify the impact of geographical factors on

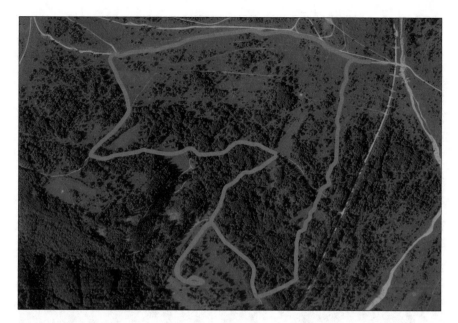

Fig. 7.47 Recorded route of Tatra 815 8 × 8 truck, 5 May 2015, MTA Libavá [*source ortophoto* CUZK (2017)]

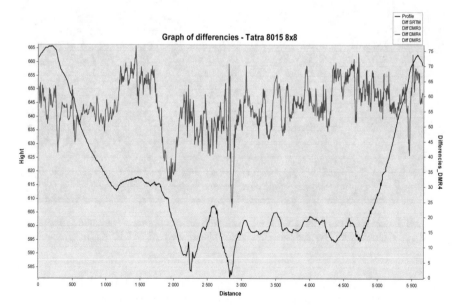

Fig. 7.48 Deviations of modelled and actual speeds of Tatra 815 8 × 8 truck in the data combination K4. A black line is an elevation route profile, a blue line are deviations of modelled and actual speed

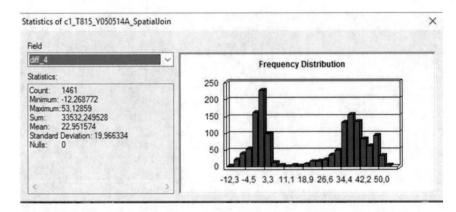

Fig. 7.49 Histogram of differences between modelled and actual speed (Tatra 815 8 × 8 truck, data combination K4)

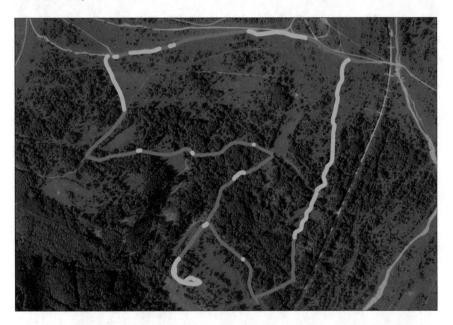

Fig. 7.50 Route sections of Tatra 815 8 × 8 truck with different surface—consolidated road (red points), open terrain with sparse shrubs (blue points) [*source ortophoto* CUZK (2017)]

the movement of vehicles using these sub-parts; remaining factors will be fixed. The impact of roads, surface and terrain relief slope was modelled as the first one.

A vehicle movement analysis was divided into two parts:

– a vehicle movement analysis on roads,
– a vehicle movement analysis in the open field.

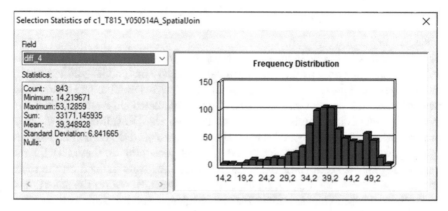

Fig. 7.51 Histogram of differences between modelled and actual speed on consolidated roads (Tatra 815 8 × 8 truck, data combination K4)

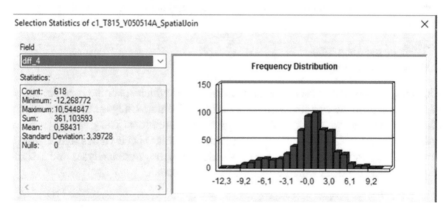

Fig. 7.52 Histogram of differences between modelled and actual speed in the open field (Tatra 815 8 × 8 truck, data combination K4)

The original deterministic model was not temporarily used for a movement analysis; however, mathematical statistics was used. In 2016 and 2017, the database of drive records in the open field was continuously supplemented with drives on consolidated roads (over 8000 km) using various types of vehicles—passenger cars, minibuses, trucks with and without loads, etc. All the measured data covering routes with total length over 10,000 km were used to find out functional or statistical dependencies between a vehicle speed and geographical conditions (surface slope, surface type, road characteristics, and road configuration).

7.5.1   Movement on Roads

The data analysis of vehicle movement records on roads showed that the speed of vehicles is affected both by the road technical characteristics and its gradient and by the road geometry as well. Therefore, the effect of curves number and characteristics on the vehicle speed was analysed using the calculated curvature parameter or measured deflection of driving direction on the road resp. Resulting from the real-time recorded data, the effect of vehicle speed was analysed depending on the road category, road width, road deflection and gradient. According to these characteristics, homogeneous sections of the same characteristics were defined on the roads and analysed using the tools of mathematical statistics. Based on the source data and measured speed values, vehicle speed dependencies on the road characteristics were calculated. Due to the impossibility to test tracked technique on the roads with the exception of unconsolidated paths, the functional dependencies were determined only for the wheeled technique. Functional dependencies were calculated in two hierarchical levels:

– Truck or passenger car,
– Particular vehicle type.

The analysis resulted in hierarchical decision tables which specified how the speed had been determined or calculated for a particular vehicle category or directly for a particular vehicle type under specific geographical conditions. The given tables were based on hierarchical schemes showing both particular speed values and assessed functional dependencies. Unlike the previous model, the entire analyses and assigning model speed values were performed in vector format.

7.5.2   Vehicle Movement Analysis in the Open Terrain

Based on the measured data analyses, mathematical models for the movement speed of both wheeled and tracked technique in the open field were also refined using statistical models. In principle, it turned out that the slope of the terrain starts affecting the vehicle movement (both wheeled and tracked) off the road starting roughly from 10°. Up to this gradient, the movement is affected mostly by the surface roughness and micro-relief shapes. If the gradient exceeds 10°, the performance parameters of a particular vehicle are already shown.

The results of analyses were again included in the decision tables and hierarchical schemes listed in the attachments, which are used to calculate the cost map of model speeds. Unlike the analyses and models of vehicle movement on the roads, these analyses are performed in a raster format with a raster cell size of 5 × 5 m.

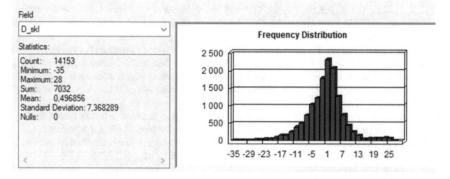

Fig. 7.53 Differences between modelled and actual speed for Tatra 815 8 × 8 truck in the area of MTA Libavá

7.5.3 *Results of the Reduced Model*

The reduced model was processed in the program system ArcGIS as well. By testing the functionality of the model using the data from which it had been derived, it was assumed that the model speed values would have higher reliability than that achieved with the complex model. The Fig. 7.53 shows the distribution of frequency variations between the model and actual speed of Tatra 815 8 × 8 truck in the area of MTA Libavá. Axis *x* shows differences between modelled and actual spead. The *y* axis shows the number of cases. The speed deviations were calculated from the routes travelled by this truck from the tests in 2015 and 2017. The recorded routes were both in the open field and on consolidated and unconsolidated roads.

The distribution of speed deviation rates (Fig. 7.53) no longer exhibits local extremes and approaches the ordinary distribution. Thus, it is obvious that the simplified model can be considered as verifiable in terms of modelling the effects of the surface slope, surface type, road characteristics and its configuration. These effects can be fixed in further development of the model, and the model can be developed by introducing other factors and analysing their effects.

References

Alvin, R. B., & Haley, P. W. (1992). *NATO reference mobility model, Edition II, NRMM II users guide* (Technical Report GL-92-19 ed.). Vickburg, USA: Department of the Army-Waterways Experimental Station, Corps of Engineers.

ArcGIS. (2015). ArcGISHelp. Retrieved from www.esri.com.

CMeS. (2017). *Meteorological dictionary (Meteorologický slovník—In Czech).* Retrieved 03 2018, from Meteorologický slovník výkladový a terminologický: http://slovnik.cmes.cz/.

CUZK. (2017). *Geoportal CUZK.* Retrieved from WMService—WMS—Ortofoto: http://geoporta l.cuzk.cz/WMS_ORTOFOTO_PUB/WMService.aspx?.

CUZK. (2018). (State Administration of Land Surveying and Cadastre, Land Survey Office) Retrieved 10 2017, from Geoportal CUZK—Access to map products and services: http://geo portal.cuzk.cz/.

D'Amico, P., Di Martino, F., & Sessa, S. (2013). A GIS as a decision support system for planning sustainable mobility in a case-study. In A. Ventre, A. Maturo, S. Hoskova-Mayerova, & J. Kacprzyk (Eds.), *Multicriteria and multiagend decision making with applications to economics and social sciences* (Studies in Fuzziness and Soft Computing ed., pp. 115–128). Berlin Heidelberg, Germany: Springer.

De Felice, F., & Petrillo, A. (2013). Decision making analysis to improve public participation in stategic energy production management. In A. Ventre, A. Maturo, S. Hoskova-Mayerova, & J. Kacprzyk (Eds.), *Multicriteria and multiagend decision making with applications to economics and social sciences* (Studies in Fuzziness and Soft Computing ed., pp. 129–142). Berlin Heidelberg, Germany: Springer Verlag.

Di Martino, F., & Sessa, S. (2011). Spatial analysis and fuzzy relation equations. *Advances in Fuzzy Systems, 2011*, p. 14, Article ID 429498.

Esri. (2013). ArcGIS User documentation. Copyright © 1995–2013 Esri.

Hofmann, A., Hoskova-Mayerova, S., & Talhofer, V. (2013). Usage of fuzzy spatial theory for modelling of terrain passability. *Advances in Fuzzy Systems, 2013*, p. 13, Article ID 506406.

Hoskova-Mayerova, S., Hubacek, M., Bekesiene, S., & Bures, M. (2017). Vehicle movement modelling possibilities for defense and crisis management. In M. Čepin &. R. Briš (Eds.), *Safety and reliability—Theory and applications* (pp. 3035–3039). London: Taylor & Francis Group.

Hošková-Mayerová, Š., Talhofer, V., & Hofmann, A. (2013). Decision-making process with respect to the reliability of geo-database. In A. G. Ventre, A. Maturo, Š. Hošková-Mayerová, & J. Kacprzyk (Eds.), *Multicriteria and multiagent decision making with applications to economics and social sciences* (Studies in Fuzziness and Soft Computing ed., pp. 179–195). Berlin Heidelberg, Germany: Springer.

Hubacek, M. (2011). Geoinformatics in support of simulators training. In *Proceedings of 19th International Conference Geography and Geoinformatics: Challenge for Practise and Education* (pp. 419–426). Brno: Masaryk University, Faculty of Education.

Hubacek, M., Almasiova, L., Brenova, M., Bures, M., & Mertova, E. (2016). Assessing quality of soil maps and possibilities of their use for computing vehicle mobility. In *Central Europe area in view of current geography* (pp. 99–110). Brno: Masarykova univerzita.

Jayakumar, P., & Dasch, J. (2016). The Next Generation NATO Reference mobility model development. *NATO Public Release documents*, 22. Retrieved from https://www.sto.nato.int/publicati ons/pages/results.aspx?k=NG%20NRMM&s=Search%20All%20STO%20Reports.

JPL, N. (2015). *Shuttle radar topography mission*. Retrieved from NASA—Jet Propulsion Laboratory: http://www2.jpl.nasa.gov/srtm/.

Kainz, W. (2007). *Fuzzy logic and GIS*. Vienna, Austria: University of Vienna.

Mitchell, A. (1999). *The ESRI guide to GIS analysis*. Redlands: ESRI Press.

MoD-GeoS. (2010). *Catalogue of the topographic objects DMU25* (7.3 ed.). Dobruska: Ministry of Defence of the Czech Republic, Geographic Service.

NDMC. (2017). *National Drought Mitigation Center*. Retrieved 03 2018, from National Drought Mitigation Center Website: http://drought.unl.edu/DroughtBasics/TypesofDrought.aspx.

Novák, P. (2000). *Soil database (Účelová databáze PŮDY, Příručka uživatele—In Czech)*. Praha: Vojenský zeměpisný ústav.

Pokonieczny, K., & Mościcka, A. (2018). The influence of the shape and size of the cell on developing military passability maps. *Cartography and Geographic Information Science*.

Rybansky, M. (2009). *Cross-country movement, the impact and evaluation of geographic factors* (1st ed.). Brno, Czech Republic: Akademické nakladatelství CERM, s.r.o. Brno.

Rybansky, M., Hofmann, A., Hubacek, M., Kovarik, V., & Talhofer, V. (2015, November). Modelling of cross-country transport in raster format. *Environmental Earth Sciences, 74*(10), 7049–7058. https://doi.org/10.1007/s12665-015-4759-y.

Rybansky, M., & Vala, M. (2009). Analysis of relief impact on transport during crisis situations. *Moravian Geographical Reports, 17*(3), 19–26.

Rybansky, M., & Vala, M. (2010). Relief impact on transport. In: *ICMT'09: International Conference on Military Technologies* (pp. 551–559). Brno: University of Defence.

Schmid, S., Galicz, E., & Reinhardt, W. (2015). WMS performance of selected SQL and NoSQL databases. In *International Conference on Military Technologies (ICMT 2015)* (pp. 311–316). Brno: University of Defence.

STANAG 2999. (2012). *Use of helicopters in land operations doctrine* (9th ed.). Brussels: NATO Standardization Agency (MAS).

Svatonova, H., & Rybansky, M. (2014). Visualization of landscape changes and threatening environmental processes using a digital landscape model. *IOP Conference Series: Earth Environment Science, 18,* 12–18. IOP science.

Talhofer, V. (2004). Digital geographic data: Potential evaluation. In *AGILE 2004, 7th Conference on Geographic Information Science, Conference Proceedings* (pp. 675 – 686). Heraclion, Crete, Greece: AGILE.

Talhofer, V., Hoskova, S., Hofmann, A., & Kratochvil, V. (2009). The system of the evaluation of integrated digital spatial data realibility. In *6th Conference on Mathematics and Physics at Technical Universities* (pp. 281–288). Brno: University of Defence.

TIFP. (2015). *Technical Institute of Fire Protection (in Czech)*. (Ministry of the Interior of the Czech Republic) Retrieved 02 2018, from Fire Rescue Service of the Czech Republic: http://www.hzscr.cz/hasicien/default.aspx.

Veregin, H. (1999). Data quality parameter. In P. A. Longley (Ed.), *Geographic information system* (2nd ed., pp. 179–189). New York: Wiley.

VGHMÚř. (2015). Katalog topografických objektů DMÚ25. *Vydání 04 2015*. Dobruška: Vojenský geografický a hydrometeorologický úřad (in Czech).

VTUPV. (2010). *The Military Technical Institute of Ground Forces*. (The Military Technical Institute) Retrieved 01 2018, from The Military Technical Institute: http://www.vtusp.cz/en/.

Zadeh, I. (1965). Fuzy sets. *Information and Control, 8,* 338–353.

Chapter 8
Conclusion

In this monograph, the team of authors attempted to show the effect of quality of digital spatial data on analyses used in decision-making processes. The team presented a methodology for evaluating the quality of digital geographic data in terms of their utility properties when solving tasks and showed practical examples of how this methodology can be applied when solving specific tasks. This methodology was accompanied by a general method of multi-criteria analysis.

Illustrated by a specific example, the team of authors introduced the whole process of using the multi-criteria analysis and the solution of variations, including the verification process of the analytical model by tests done right in the field. For this example, they used the model solution of Cross_Country_Movement for military vehicles.

The authors are aware that the proposed methods and procedures are not the only way to evaluate the quality of digital data and their utility properties. Nevertheless, they believe that the suggested model of evaluation is universal especially because of its orientation on user needs. The proposed model can be further developed, adding or reducing the number of criteria used, setting different scales etc. However, the advantage of the model is its independence from data formats, used technology, software tools etc.

The authors hope that the monograph will contribute to critical approach to the data used by the users and to a responsible approach while producing the data by the publishers.

© Springer International Publishing AG, part of Springer Nature 2019   189
V. Talhofer et al., *Quality of Spatial Data in Command and Control System*, Studies
in Systems, Decision and Control 168, https://doi.org/10.1007/978-3-319-94562-0_8

Summary

The current way of management of activities of intervention mobile units in a given area cannot work without a well-thought way of command and control. Command and control may be generally defined as a system consisting of personal, technical, and legislative components created to provide coordination and commanding of operational units with the aim to complete required goals of an operation. A significant part of command and control is played by the decision-making process, during which an individual or a team chooses the best solution of the all possible ones. Depending on the task, facts about the given task are found, as well as connections with solution of other tasks, factors that influence the given task, and preferences that need to be considered before making the decision. One of the main factors that influence the fulfilment of given tasks are geographic conditions in the given area. To understand this condition, paper maps or digital geographic data are mainly used.

Before digital geographic data became widely used, commanders and staff got geographic information about the area of activity first of all from paper maps. Present command and control systems are characterised by the wide usage of communication and information technologies which enable commanders and staff to use digital geographic data as a primary source for evaluation of the area of activity from the geographic point of view. The efficiency of decision-making processes using maps and digital geographic data has always been significantly influenced by their quality. Quality prepared and up-to-date maps and used data enable to get detailed information about the area of activity that is necessary for making optimal decisions. For the needs of decision-making processes, there are maps and published digital geographic data that correspond to the given needs, e.g. units of integrated rescue systems or armed forces, and they are usually produced by state organizations that guarantee its constant quality, reliability and regular updating.

For the overall assessment of the quality of digital geographic data, it is possible to use the general view of the utility value of a product based on general principles of multi-criteria analysis. Multi-criteria analysis can be used to evaluate functionality of digital data or to analyse geographic conditions in a given space. The book describes the use of multi-criteria analysis for improving the utility qualities of digital geographic data. Multivariate techniques are quite frequent and complex. Mathematically they are based on a multi-criterial analysis for decision-making processes.

The strength of digital geographic data together with tools for working with them lies in the possibility to create a spatial geographical analysis in which the influences of geographical environment on the intended or realized activity are evaluated. General approaches of multi-criteria analyses may be used in many areas of decision-making processes. The book shows its applications for digital geographic data and the evaluation of their functionality. The text is also dedicated to the use of multi-criteria analyses for solutions of geo-process models. To illustrate the complexity of the method of solution, an example of modelling of passing through terrain with off-road vehicles was chosen.

Index

© Springer International Publishing AG, part of Springer Nature 2019
V. Talhofer et al., *Quality of Spatial Data in Command and Control System*, Studies
in Systems, Decision and Control 168, https://doi.org/10.1007/978-3-319-94562-0

Printed in the United States
By Bookmasters